AF718733

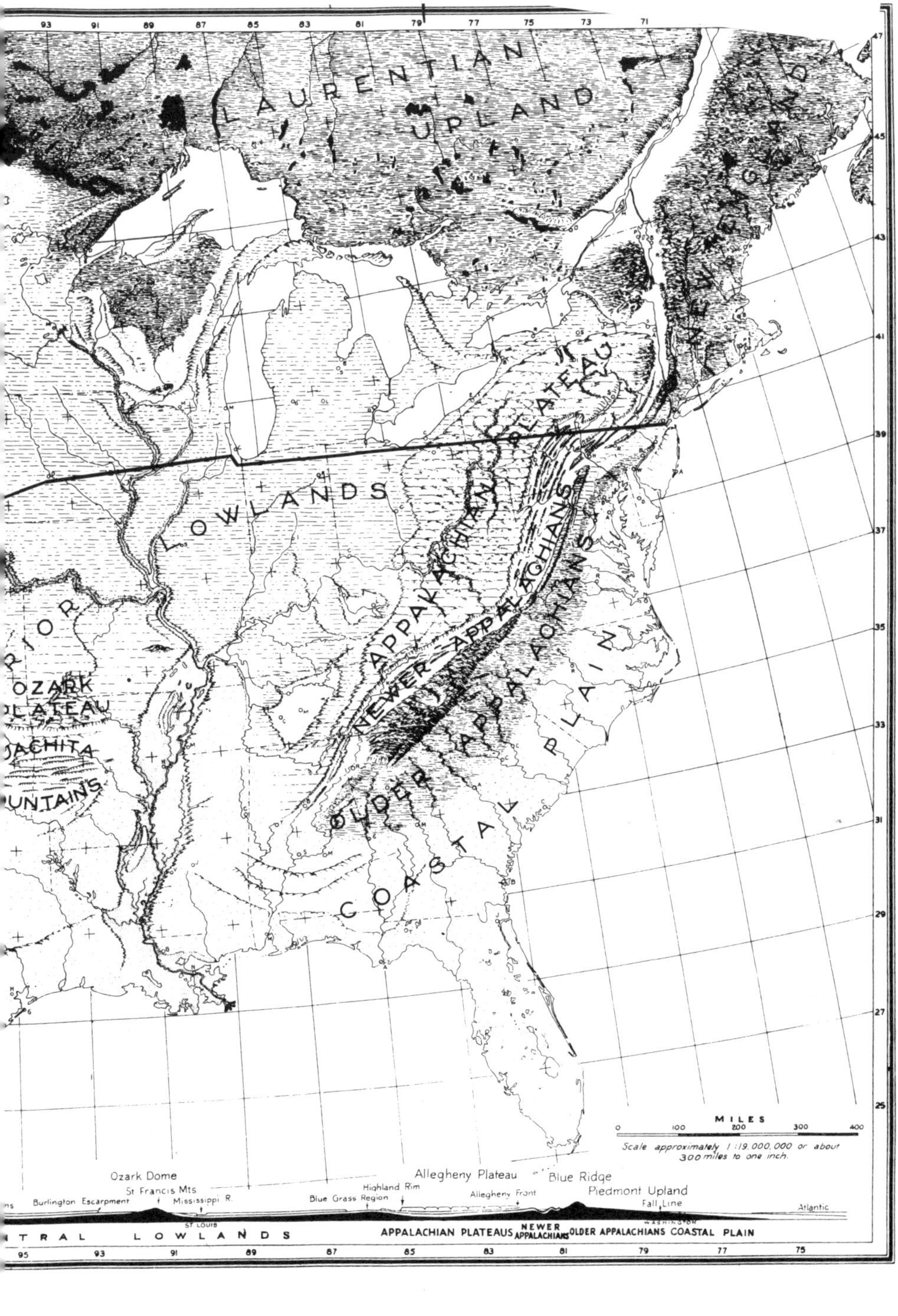

LAURENTIAN UPLAND
NEW ENGLAND
PLATEAU
LOWLANDS
APPALACHIAN
NEWER APPALACHIANS
OLDER APPALACHIANS
COASTAL PLAIN
OZARK PLATEAU
OUACHITA MOUNTAINS
MILES
0 100 200 300 400
Scale approximately 1:19,000,000 or about 300 miles to one inch.
Ozark Dome
St. Francis Mts
Burlington Escarpment
Mississippi R.
ST LOUIS
Highland Rim
Blue Grass Region
Allegheny Plateau
Allegheny Front
Blue Ridge
Piedmont Upland
Fall Line
WASHINGTON
Atlantic
TRAL LOWLANDS
APPALACHIAN PLATEAUS
NEWER APPALACHIANS
OLDER APPALACHIANS
COASTAL PLAIN

AIRWAYS OF AMERICA

KENNIKAT PRESS SCHOLARLY REPRINTS

Dr. Ralph Adams Brown, Senior Editor

Series on

MAN AND HIS ENVIRONMENT

Under the General Editorial Supervision of

Dr. Roger C. Heppell

Professor of Geography, State University of New York

(Photo by United Air Lines)

The Crest of the Wasatch Mountains, Utah.

AIRWAYS OF AMERICA

THE UNITED AIR LINES

A Geological and Geographical Description of the Route from New York to Chicago and San Francisco

By

A. K. LOBECK, A.M., Ph.D.

Professor of Geology in Columbia University

Author of "The Midland Trail", "The Mammoth Cave National Park", "A Popular Guide to Allegany State Park", "The Superb Position of New York City as a Center for the Study of Physiography", "The Physiography of Porto Rico", "Physiographic Diagram of the United States", "Europe", etc.

With maps and illustrations

KENNIKAT PRESS

Port Washington, N. Y./London

AIRWAYS OF AMERICA

First published in 1933
Reissued in 1970 by Kennikat Press
Library of Congress Catalog Card No: 78-113287
ISBN 0-8046-1323-0

Manufactured by Taylor Publishing Company Dallas, Texas

KENNIKAT SERIES ON MAN AND HIS ENVIRONMENT

CONTENTS

SUPPLEMENT

ACKNOWLEDGMENTS

A sincere expression of appreciation is due to the personnel of United Air Lines which operates the air mail-express-passenger planes on the New York-Chicago-Pacific Coast route, especially to Mr. V. P. Conroy, District Traffic Manager at New York, to whose attention the plan of an air guide was first brought; to Mr. R. W. Ireland, of the General Traffic Department; and to Mr. Harold Crary, Manager of the Advertising Department. These gentlemen together arranged for the passage both ways across the continent during the month of June 1931, at which time the notes and pictures for this guide were taken. At numerous points along the route the various officials of United Air Lines exhibited the most agreeable kind of courtesy which doubtless was in no way different from that extended to all of their patrons. It is a privilege to mention the name of Professor C. P. Berkey of Columbia University for the enthusiasm which he displayed toward the undertaking and for his help in bringing it to the attention of the air transport company. The Kemp Memorial Fund, established in memory of Professor James Furman Kemp, for many years head of the Geology Department of Columbia University, is wholly responsible for financing the publication of this book.

In the organization of the guide itself, Prof. Guy-Harold Smith of Ohio State University showed indefatigable industry in preparing the thirty-nine route maps which represent as in a bird's-eye view a complete panorama across the United States. The efficient manner in which Professor Smith and the staff of the United States Geological Survey together worked out details of cartography and lettering was of inestimable help. The cordial relations which Mr. H. G. Ferguson of the Survey maintained with Professor Smith eased the work in many ways. Mr. Walter J. Roth, Senior Agricultural Economist of the Department of Agriculture, helped to secure information concerning the agriculture of the regions flown over. Mr. C. P. Barnes, Assistant Agricultural Economist, was generous in permitting the reproduction of a strip of his map showing the agricultural provinces of the United States, and in providing brief descriptions of the different units. Several of the agricultural experts in the different states crossed by the air route assisted by furnishing interesting accounts of the conditions to be observed while flying across their states. It is impracticable to mention the names of all of those who were thus helpful, but nevertheless it is a pleasure to record the kindness of Mr. J. A. Hill, Dean of the College of Agriculture of the University of Wyoming, Mr. R. L. Adams, Professor of Farm Management of the University of California, and Mr. F. B. Heddley, Chief of the Department of Farm Development of the University of Nevada, for very useful information which has been embodied in the account of those states.

The manuscript was critically read by the author's brothers, Albert E. Lobeck and E. Quintus Lobeck, both of whom offered many happy suggestions tending to make it more interesting for the general reader.

(*Photo by Daily Pantagraph, Bloomington, Ill.*)

FIG. 1. The undissected till plains of Illinois with a very regular rectangular arrangement of farms and roads, taken from an elevation of 500 feet. The large light-colored field in the foreground covers about 16 acres.

INTRODUCTION

Diversity of Landscapes. The traveller who flies from coast to coast across the United States looks down upon a varied panorama of landscape forms. Below him pass almost every type of feature known to the geologist. There are *plains* with their many aspects; simple plains almost devoid of streams, vast in their extent, reaching as they do in all directions to the distant horizon. These are the Great Plains of Nebraska and Wyoming. There are also the *till plains* of Ohio and Illinois, equally flat but less extensive, resembling gigantic checkerboards from the rectangular pattern of their farms and roads. (Fig. 1) In Iowa, on the other hand, the *prairie plains* with their multitudinous stream courses produce a rolling landscape, pleasing in its diversity.

Elsewhere, as in western Pennsylvania, the land is high and the streams flow in deep gorges, almost canyon-like in their appearance. Bold and savage escarpments descend from the upland surface, revealing here and there in their deeply wooded recesses the suggestion of a tumbling stream. Such areas are termed *plateaus.* They are called *young plateaus* if trenched by a few widely spaced canyons as along part of the Colorado-Wyoming boundary; *mature plateaus* if intricately dissected by many water courses as in Pennsylvania; or *old plateaus* if dominated by *mesas* and *tablelands,* as they are in central Wyoming. These represent the last remnants of extensive areas now almost completely worn away by the agents of erosion.

But it is in the *mountains* that the student of geomorphology takes the keenest interest. A cross-section of the United States from New York to San Francisco reveals to the geologist the widest range of geological structures. These structures in turn affect the topography, each in its own peculiar way. Some structures are simple as in the case of a plain bowed up slightly to form a *dome.* In central Wyoming several domes may be seen. In some cases they are almost intact, but most of them, like the Baxter Uplift at Rock Springs, have had their entire crests worn away so as to reveal a series of concentric outcrops where the deeper beds have been exposed. To the traveller viewing this from the air, this is ân object lesson never to be forgotten.

Other mountains, somewhat more involved in their structure, have resulted from a regular bending of sedimentary formations so that gigantic wrinkles or *folded mountains* were produced in the earth's crust. As usual such areas have been profoundly worn away by streams and other forces of destruction, and now only the roots of the original mountains remain, forming long ridges across the country. No region in the world can rival the belt of the Folded Appalachians which traverses eastern Pennsylvania. Only the true lover of land forms can appreciate the joy of the geologist who is privileged to look down upon such a landscape, and see, revealed in its topography, the internal structure of the rocks which make up the earth's crust.

But there are still other mountains of even greater interest. The *block mountains* of the Great Basin region in Nevada are not yet through with their growing pains. These ranges from year to year are heaved up along the fractures or rifts which originally caused them. In several places in Nevada the air route passes directly over these recent breaks, one of which occurred in 1915 and produced a conspicuous scarp at the foot of the mountains.

The highest and probably the most extensive and most numerous of the mountain areas of the United States are those which are termed *complex mountains.* The Sierra Nevada, the Rocky Mountains, the Wasatch, and the Highlands of New Jersey all come in this category. In New Jersey the depth of soil cover and the richness of the vegetation effectively conceal the rock masses. But in the western mountains the rocks are laid bare to view, red granite in the Laramie Range of Wyoming, white

(*Photo by United Air Lines*)

Fig. 2. The Front of the Wasatch Range, at Salt Lake City.

granite in the Sierra, and complexly folded and much disturbed formations in the Wasatch and other ranges of the Great Basin.

The long sprawling spurs of complex mountain ranges descend from the higher summits, gradually dying out as foothills on the adjacent plains like the outspread paw of a huge beast. Some complex mountain areas, notably the Wasatch and the basin ranges of Nevada, are bewildering, tumbling masses of sharp-crested ridges and pointed peaks which in the early morning and late evening hours cast long and weird shadows far over the surrounding country. (Fig. 2)

Along the air route no great and active volcanoes are encountered, but several small and very perfect *cinder cones* may be seen in western Nevada. The *craters* of these cones are still intact or only slightly breached by stream erosion, and in one or two instances an easily recognized *lava flow* extends from the base of the cone for a mile or so over the adjacent land. The location of these small volcanoes on a fracture zone along which the molten rock has found a way to the surface is also interesting and significant. Conspicuous white or delicately hued *hot spring deposits* like the mammoth terraces of Yellowstone Park are likewise to be observed along fault lines at the base of several of the block mountains of the Great Basin.

To recount the bewildering variety of detailed forms resulting from the activities of *streams*, of *glaciers*, of *waves*, and even of the *wind* would resemble an index to a treatise on geology. To observe these things in their proper setting is, however, an experience of the greatest interest. Take at random the marvelous festoons of *glacial moraines* in northern Illinois and Indiana, revealing themselves by the many *lakes* and *ponds*, the irregular roads and the wood lots of the farms or even by gravel pits, rather than by their actual topography which is too subdued to be detected from aloft; or the *old beaches* of Lake Michigan running for miles through the railroad yards and the factory sites on the outskirts of Chicago; or the *shorelines* of ancient Lake Bonneville on the flanks of the mountains near Great Salt Lake looking like artificial roadways or railroad embankments; or the *braided channel* of the South Platte River in central Nebraska with its ribbon of irrigated land on either side; the sea of *sand hills* covering thousands of square miles in the western part of that same state; the long *alluvial plains* sloping gently away from all the ranges of the Great Basin region; and the *embayed shore* of Lake Erie at Sandusky Bay. These features and many others of equal significance would each one suffice as the key wherewith to unlock a long and interesting story.

In the belief that readers of this guide would relish a systematic account of the topography and structure of the United States, there has been added in the latter part of the book a chapter on the physiographic provinces of the United States traversed by the air route. And it is suggested that this chapter might be read with profit in a leisurely and studious way before undertaking the more detailed descriptions of the route itself. Plates I and II in the pocket show the physiographic provinces and the geological formations crossed by the air route.

Many Kinds of Agriculture. Lest these remarks seem to betray too strongly an interest in the landscapes with no sympathy for those who dwell therein, it is well to emphasize next an entirely different aspect of this journey. The traveller by air will probably give much thought and attention to the never ending panorama of farms and fields which hour after hour unrolls itself before him. (Fig. 3) From the Atlantic to the Pacific coast it is hardly possible to glance from the window of the plane without beholding some form of agricultural activity. The numerous *truck gardens* and *vegetable plots* near New York give way gradually to the *open farms* of western New Jersey and Pennsylvania and then come the *orchards* of the Allegheny Plateau, followed by the *dairy* and *live-stock* areas of the Middle West. There, even the color of the cows takes on some significance, . near the large centers like Chicago, the black and white Holstein breeds, famous for their yield of milk, dominate the

(Photo by Daily Pantagraph, Bloomington, Ill.)

FIG. 3. Combination dairy and grain farm at Forrest, Illinois.

landscape; but in western Illinois and in Iowa the production of beef is more important and the brown Durham and other types with similar tawny tones seem to be in greater vogue. Beyond the Missouri River *irrigation* becomes important. The valley of the Platte River winds like a ribbon of green across the dry plains of Nebraska displaying in a very impressive way the contrast between irrigation and *dry farming*. The traveller by train sees only the verdant fields of the valley floor and were he not otherwise informed might conclude that this is representative of the whole state. Still farther west on the open *ranges* and in the park-like areas of the higher mountains the *grazing* of sheep and cattle becomes dominant.

This has a seasonal aspect which is interesting. During the summer the arid land of Wyoming constituting the Red Desert is parched and grassless. For a hundred miles there is no sign of animal or habitation of man. But in winter and spring the vegetation takes on new life and then herdsmen with countless thousands of sheep may be seen with their canvas covered wagons which serve as their homes. And, finally, in the sunny land of the farthest west there are the rich orchards of the Great California Valley, one of the garden spots of the world. (Fig. 4)

Students of agriculture recognize forty-five distinct regions between New York and San Francisco, each one characterized by its own peculiarities of soil and crops. Little wonder that the air traveller finds much to interest him in the behavior of his fellow men who live on the soil. The kinds of crops, the livestock, the size and shape of farms and farm buildings, vary widely from place to place. These variations are sometimes puzzling for there are many factors which affect the farmer's choice of action and to some extent determine those things which he will do.

Fortunately the sympathetic guidance of the U. S. Department of Agriculture has made possible the introduction of a chapter which describes the forty-five agricultural units to be observed from the air. A map showing the extent of these areas constitutes Plate III in the pocket of this book.

Great Contrasts in Natural Vegetation. In the third place, an air journey across the continent brings rapidly before the observer a succession of regions distinguished because of their natural vegetation. In the eastern half of the United States much of the original forest has disappeared under the hand of man, but every ridge and rugged bit of land from New York westward into Ohio and Indiana is still clothed with a *hardwood forest*. One of the wildest and least developed parts of the eastern United States is in western Pennsylvania where the forest covers many square miles of the plateau, and is crossed only by little woods roads whose deserted appearance is quite in contrast with the busy highways in the adjacent cultivated valleys. In the northeast this forest has a large percentage of soft wood or coniferous trees but westward in the Mississippi valley it is a part of the *southern hardwood forest* with oak and hickory as the dominant types. Then westward of this come the *prairie grasslands* of Illinois and Iowa, now almost everywhere turned into agricultural land. These give way in the Great Plains, with their drier climate, to a region of *short grasses*, and eventually to the *bunch grassland* of the far west. It is in these sparsely covered soils that the prairie dogs make their homes which to the air traveller resemble miniature shell marks on a battle-scarred field in France. The *western pine forest* covers the several ranges of the Rockies, the Wasatch and the Sierra Nevada. Usually open and park like, and especially in the higher altitudes with luscious meadows spreading through the scattered groves of trees, these forests largely untouched by the hand of man may be observed in all their natural beauty at scattered intervals for a thousand miles from Wyoming westward.

The ranges in the arid lands of the Great Basin support a desert type of forest cover made up largely of *piñon* and *juniper*, trees of a scrawny nature able to survive on steep and rocky slopes devoid of soil.

FIG. 5. Cumulus clouds over the plains of eastern Nebraska, taken from an elevation about 8,000 feet above the ground.

The Operation of Airlines. Chapter V may prove of interest to those inquisitive persons, in this case more likely the men than the ladies, who want to know what makes the wheels go round. The operation of an airline is a big and complicated matter most of which will escape the traveller's attention unless he makes it his business to find out something about it.

The selection, training, and experience of *pilots* is perhaps the topic of prime interest. The air traveller knows that pilots must have proven experience, but he perhaps does not always realize that they must also maintain a high standard of personal conduct and physical condition at all times. Pilots fly usually on alternate days, on a four hour run. This means for example, that one set of pilots covers the

FIG. 6. Cumulus clouds over the Allegheny Plateau, taken from an elevation of almost 9,000 feet above the ground.

FIG. 7. The interior of a transport plane. Luncheon is being served aloft. At night the passengers are made comfortable by reclining their chairs. Blankets and pillows are also provided.

territory between New York and Cleveland, another set between Cleveland and Chicago. Chicago to Omaha, Omaha to Cheyenne, and Cheyenne to Salt Lake City are other divisions of the route. The longest run of all is the run from Salt Lake City to San Francisco, almost a seven-hour trip. In the course of a few years a pilot manages to cover his run hundreds of times and in several instances a thousand and even two thousand trips have been credited to one man. During this period the terrain over which he flies becomes known to him in all of its intimate detail and in all of its various aspects from season to season. The average flying experience of the 150 transport pilots serving the United Air Lines is 4000 hours. Included among these fliers are nine, each of whom has flown in excess of a million miles and many of them have flown over 5000 hours. A substantial number of them have 10,000 hours to their credit.

When the traveller first makes the acquaintance of the *stewardess* on an air liner he may not realize at first what a superior type of person she is. Her ostensible duties involve the collection of the passengers' tickets, the checking of baggage, the serving of meals en route and looking after the comfort of all. (Fig. 7) But she is also a trained nurse, and is a person of integrity, character, and good judgment, able to cope with almost any emergency.

The traveller by day only occasionally catches glimpses of the numerous *beacons* which mark the airway across the country but at night the flashing lights, spaced at 10 mile intervals are easily seen, sometimes five or six of them at a time in either direction, marking the route for a hundred miles or more over the plains. It is then that he becomes interested in the navigation of planes and in the several *radio* devices for signalling, for guiding and for voice communication. It is hoped that the chapter on airways and avigation will to some extent satisfy this interest and that the maps showing the radio stations and beacons along the air route will prove of value to some of those who use this guide in actual flight.

Observations on a Flight. The sensation of flying is not what most people imagine. There is no feeling of dizziness such as some people experience when looking over the side of a tall building. Because there is no direct connection with the ground the person in the air does not have any means of measuring the height nor any tangible evidence that his position aloft must be sustained in some manner. The sensation is not unlike that gained in looking over the side of a high suspension bridge from its middle part. There is perfect calmness, possibly a feeling of exhilaration, never one of fear.

As the plane gains speed for the take-off the sense of motion is of course very great indeed. Just at the moment of leaving the ground the plane is probably moving considerably faster than most people ever travel by train or automobile and the nearness of objects renders this speed easily appreciated.

After an elevation of several thousand feet has been attained the landscape loses its relief and takes on the aspect of a map. (Fig. 8) This loss is perhaps serious to the person who expects to be impressed with the majesty of serrate peaks and the savage boldness of cliffs and precipices. But it is more than compensated by the grandeur of the panorama and the vast expanse of territory which a single view encompasses. From an altitude of 10,000 feet, for example, an area the size of New York State comes within the range of vision. When the observer's position lies below the highest summit (Fig. 9), then the profiles of mountains and ridge tops give a reality to the scene which a more lofty point of observation makes impossible. Such nearer views are, therefore, likely to be more impressive and satisfying. To be properly awed and inspired it is necessary that one be able to make some comparison between himself and the landscape.

If it is true that landscapes lose their third dimensional aspect when seen from a considerable height, how can the observer supply this defect and thus gain some

FIG. 8. The rolling plateau country of northern Ohio, near Cleveland, from an elevation of 2,000 feet.

appreciation of the relief? This he must do by indirect means. The influence which the topography exerts upon types of vegetation and upon the activities of man renders it possible for the person in the air to interpret the phenomena which he observes in terms of topographic forms. For instance, the deeper green of forested areas may be taken as an indication of topography more rugged than where cultivated fields exist. The morainal belts of Indiana and Illinois reveal themselves not by their topography. Their hummocky surface is lost to the observer, even at a slight altitude, but the patches of woods, the lack of cultivated fields, the grazing land, often with flocks of sheep, the occasional lakes, the irregularity of the roads, all of these things betoken a rougher landscape than the open country on either side with its straight roads and large farms. In other words, it is often the geographical rather

FIG. 9. The eastern slopes of the Wasatch in northern Utah, viewed through the window of a tri-motored plane, flying below the mountain tops.

FIG. 10. Bluffs of the Green River formation near Rock Springs, Wyoming. View from a west-bound plane flying at a low altitude.

than the physiographic aspect of a region which commands attention. Properly interpreted, these phenomena become the means by which the geomorphology can be deciphered. Each gains significance from the other.

When a flight is made at an elevation of a thousand feet or less above the ground, then many of the details of topography become apparent. (Fig. 10) Owing to the fact that most scheduled flights toward the west are made at a low elevation to avoid the strong westerly winds prevailing at high altitudes, it is obvious that much more

FIG. 11. The Loup River in eastern Nebraska. View from an eastbound plane flying at a height of a mile above the earth's surface.

is likely to be seen by the passenger going west than coming east. Eastbound flights are often made at elevations of a mile above the earth's surface, which is too high for the detection of many details. (Fig. 11) Those who travel by air with the intent of seeing the country will gain most from the westbound journey. The text of this guide is therefore arranged for the traveller going from New York to San Francisco, though it can be used also without great inconvenience in the other direction.

Maps. At the end of the book there is a chapter of general hints, suggestions and observations, which might prove of interest to a person who is actually planning a trip by air across the country. The question of maps looms up rather importantly. Some travellers like to take maps along on their trips and identify the places over which they ride. Most people are well satisfied with the most general type of map or log of the course but there are a few who are interested in details not depicted on small scale maps. A selected list of the U. S. Geological Survey topographic sheets is therefore described. Obviously it is impracticable to carry a complete set of the large scale topographic sheets on a trip nor would it be possible to describe them all in this book. A selection of thirty-six maps has therefore been made and a brief note on each one points out the features of especial interest. Carried to the extreme, this description of maps could be made a veritable text-book on geology and physiography, for which the United States would serve as a comprehensive outdoor laboratory wherein every type and every form would be found illustrated. Plate VI is an index map giving the names of all the topographic sheets crossed by the air route.

References. The list of reading references with which this book ends might easily have been multiplied several times. Like all similar lists it represents simply those books and articles which have proved useful or have come to the attention of the compiler of the list. Perhaps it will be helpful to the student not familiar with geological fields of thought. Any serious student will know how to delve into that particular branch of the subject which happens to interest him.

The lover of the past will know that Hamlin Garland's "A Son of the Middle Border" gives a vivid picture of pioneer days in the Prairie Plains country; and that Owen Wister's "Virginian" does the same for Wyoming. Francis Parkman's "Oregon Trail" might perhaps be singled out from all others as the book most likely to provide the proper historical background for a visit to the great open spaces beyond the Missouri.

The route from New York to San Francisco has been divided into thirty-nine sections, for each one of which a Route Map, or bird's-eye diagram has been prepared. The index to these maps forms Plate I.

The descriptions of the thirty-nine route maps are intended to be read during flight across the region. They are, therefore, necessarily rather brief and matter-of-fact, and to some extent they presuppose an acquaintance with the information set forth in the supplementary chapters.

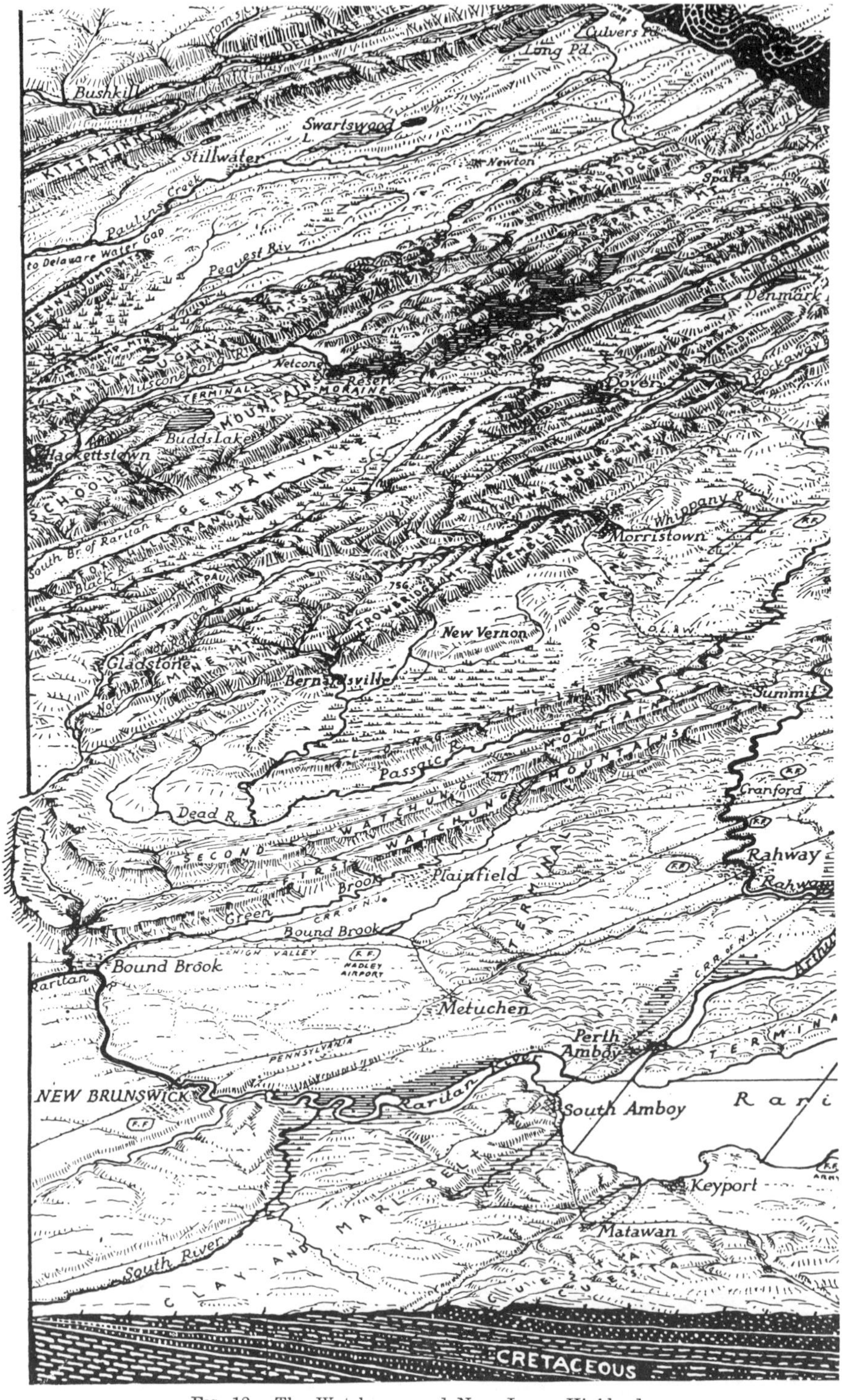

Fig. 12. The Watchungs and New Jersey Highlands.

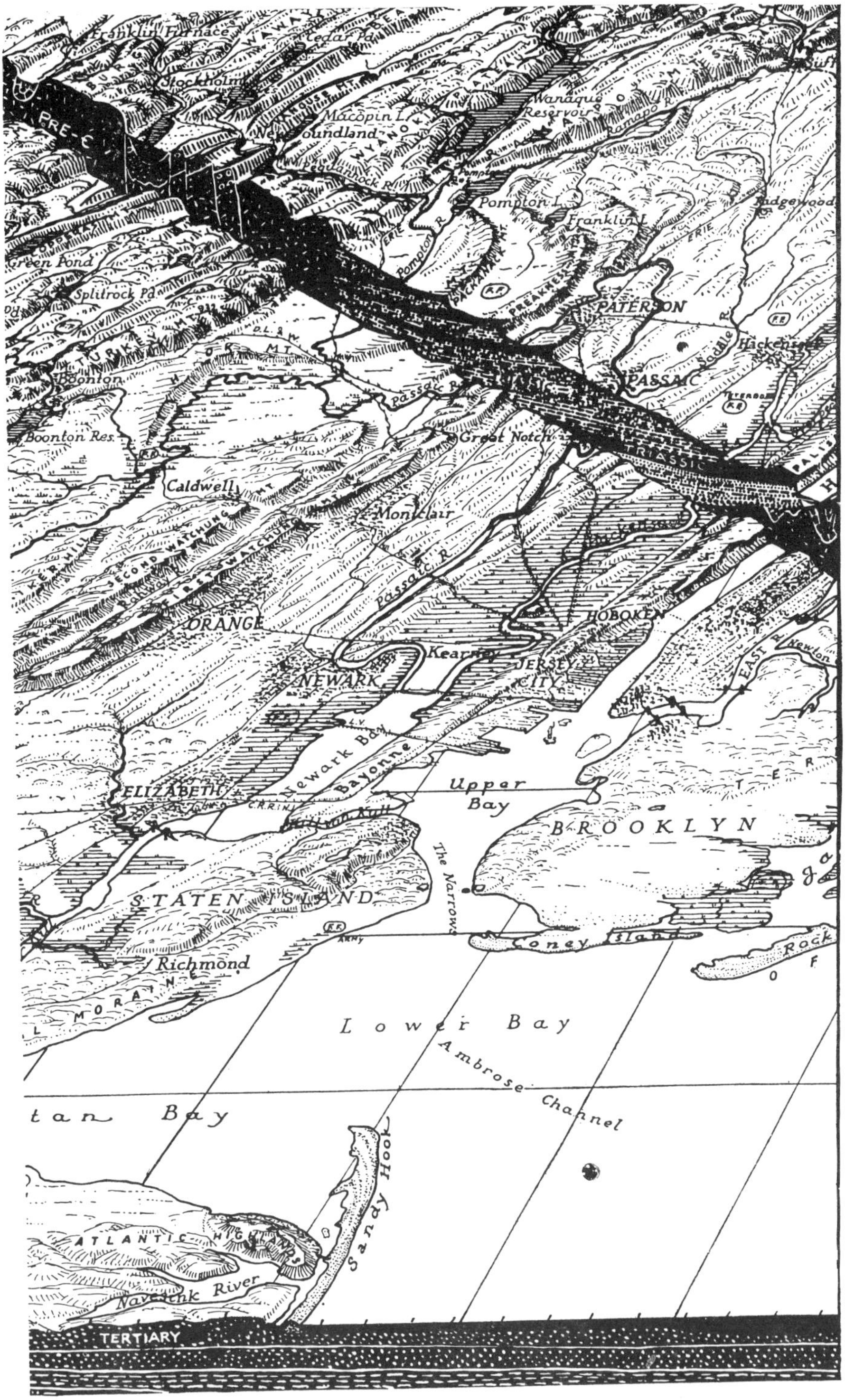

FIG. 13. The Metropolitan area near Newark.

THE ROUTE IN DETAIL

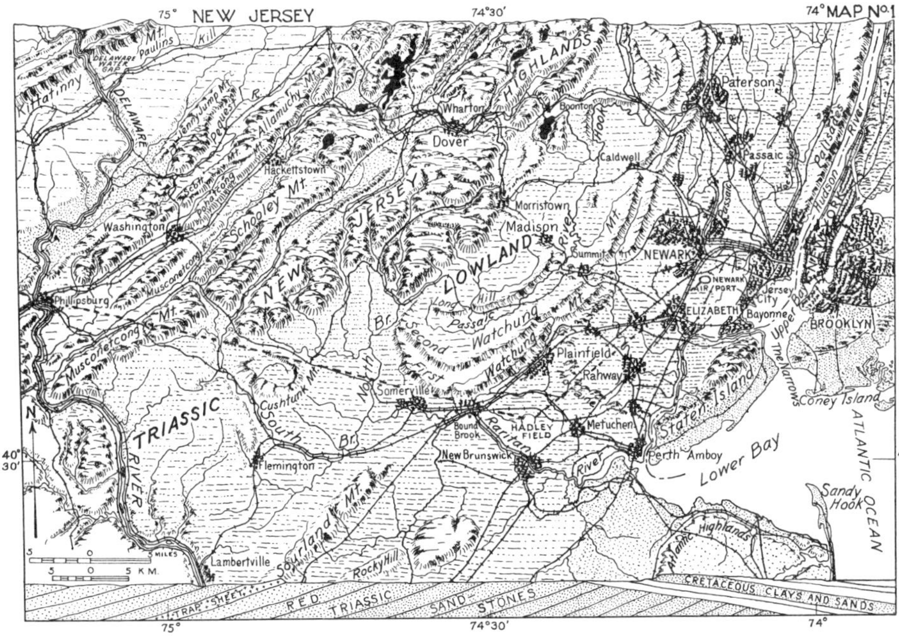

Route Map No. 1

The scale of the Route Maps is 1 : 750,000. The air route is shown by the dotted line, but this is only approximate as the actual line of flight varies in places as much as ten miles on either side.

NEW YORK TO EASTON

The Newark Region. The Newark airport occupies filled-in meadow land close to the point where the Passaic River enters Newark Bay. The Passaic and Hackensack Rivers, as well as the Raritan River farther south, are tidal estuaries. The margin of the Triassic Lowland has here been submerged.

From the airport the city of Newark stands out conspicuously two or three miles to the north, its tall buildings dominating this part of the Triassic Lowland. They can be seen from far distances in all directions. Just beyond Newark rises the forest-clad wall of the First Watchung Ridge, with an elevation of almost 500 feet above the plain on which Newark stands. Its even crest forms the western skyline for many miles.

To the east, over the low, plunging end of the Palisades, the towering buildings of downtown New York may be descried, piercing the sky. Being almost ten miles away, they are visible only in clear weather.

The Estuaries around New York City. Immediately after the take-off the vast expanse of the Triassic Lowland opens up to the north and extends as far as the eye can reach. Its grassy flats and sparkling lagoons offer a contrast to the inhabited areas on either side. If possible, the immediate glance had best be toward the east, for these features will rapidly recede from view. The long gentle back slope of the Palisades is readily discernible, its southern part entirely built upon by parts of Jersey City and Hoboken, but farther north supporting a continuous growth of dense forest. The Hudson River may for a moment come into view, separating the Palisades from the great metropolis on the east. If only for a twinkling now, one quick view should take in the scene toward the southeast and note should be made of those splendid embayments which account for the marvelous harbor facilities of New York and its environs.

Newark Bay, Upper New York Bay splitting northward into the Hudson and the East River, The Narrows, Gravesend Bay, and the Lower Bay, seen far away over Staten Island, to mention only a few of them, account for some of the 600 miles of water front which the metropolitan area possesses (Fig. 13).

And almost immediately, too, there will be a clear impression of the numerous railways and roads threading their way across the lowlands to converge upon New York.

The Atlantic Highlands. Far away to the south, and visible only in fairly clear weather, is the inface of the cuesta[1] which constitutes the most conspicuous feature of the New Jersey Coastal Plain. It is most prominently represented in the Atlantic Highlands. Farther west it becomes more subdued and swings sharply toward the south, with the result that it rapidly disappears from view.

The Watchung Ridges. In the first three or four minutes after leaving the Newark airport the observer rides over a small part of the Triassic Lowland which has been glaciated. He then crosses the terminal moraine just before the First Watchung Ridge is reached. Although the topography of the moraine may not be particularly striking at the height from which it is seen, nevertheless its course is suggested by the more abundant trees, and the fewer cultivated fields. West of the moraine and outside of the glaciated area, there is a belt of outwash plain.[2] This serves as a site for Plainfield, whence its name, and also for Hadley Airport, which is crossed by most of the air routes from the New York region to the west and south.

Soon the traveller finds himself looking down upon the two Watchung Ridges. As height is gained, the country seems to flatten out below, but the deeper green of the forest cover clearly emphasizes the position of the ridges.

The contrast between the steep eastern slope and the long gentle dip slope to the west is noteworthy, although both slopes are forested. The eastern face of the Watchungs is not a bold escarpment or cliff like the Palisades, although the actual height is about the same. Between the two Watchung Ridges and also back of the second ridge the lowland areas are everywhere under cultivation. In the broad plain back of Second Ridge the sinuous form of Long Hill, with its forest cover, appears like a winding snake. Toward the north and west the Lowland is hemmed in by the blue wall of the New Jersey Highlands whose even-topped skyline marks the level of the so-called Schooley peneplain.[3]

Occasionally when clouds are forming and moisture-laden winds are being driven inward against the Watchungs and the Highlands, these ranges will each be sur-

[1] A "cuesta" (from the Spanish) is a ridge or mountain, steep on one side and gently sloping on the other. The coastal plain cuesta has its steep side or inface toward the continent and its gentle back slope toward the ocean.

[2] The terminal moraine is a belt of low hills, most of which are less than 50 feet in height, deposited by the melting ice sheet at the end of glacial time. The "outwash plain" is a flat stretch of sand, often many square miles in area, spread in front of the moraine by the outpouring water.

[3] A peneplain is a more or less level region which has been formed by the wearing down of once mountainous areas to a lowland or plain. Such a peneplain may be raised again to form an upland.

When this happens valleys are eroded below the peneplain surface, and the country becomes rugged.

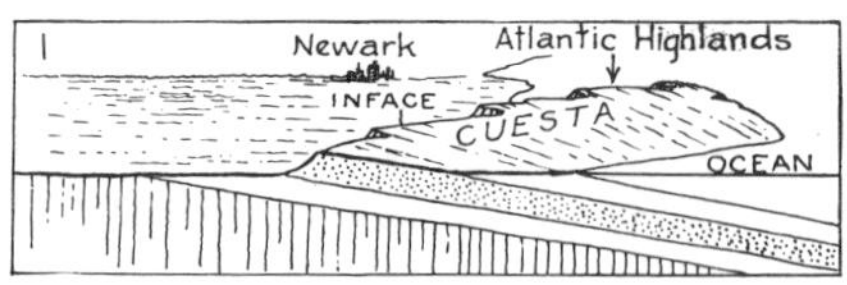

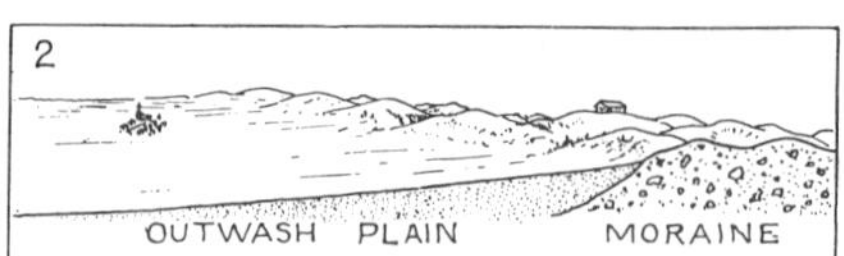

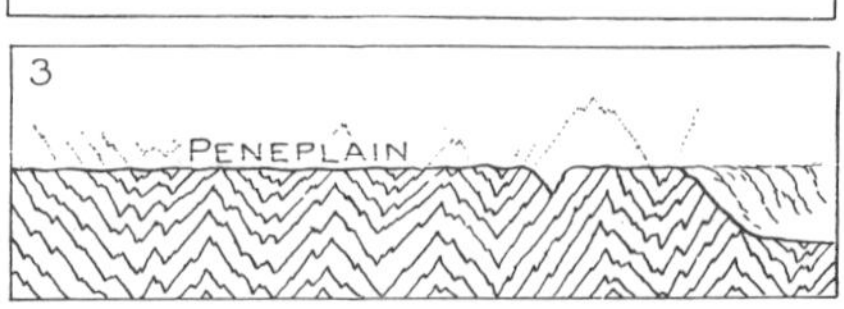

mounted by a bank of mist formed by the rising air currents. In fact, throughout this part of the state the clouds may correspond very closely with the hilly topography below.

The First and Second Watchung Mountains having been crossed, it will be well to glance to the south and observe the bow-shaped form of these trap ridges (Fig. 12), for in a moment the plane will pass over them again in reverse order. The curved form is due to the basin structure of the Triassic Lowland, this being the southern end of a broad syncline,[4] a syncline, however, which is cut off on its western side by the Highland Fault. The abrupt wall of the Highlands is a fault-line scarp[5] due to erosion along this displacement. At several points quarries may be seen in the Watchung Ridges. This trap rock is an important road-building material throughout the New York City region.

The course which has just been described lies several miles north of the lighted and marked airway which skirts around the southern end of the Watchung Ridges. This direct route westward is the one likely to be followed by the air traveller who goes by daylight and in good weather.

The New Jersey Highlands. After leaving the Watchungs behind and just before coming to Cushtunk Mountain, the route runs practically along the boundary line between the Triassic Lowland with the red sandstone on the south, and the crystalline area with its buff-colored soils on the north. The line of contact can be observed from the plane, cutting across some of the ploughed fields which are half red and half buff in color.

Cushtunk Mountain, just south of the town of Lebanon, is a horseshoe-shaped trap ridge of unique form. It appears to be due to warping, and because of its steep inner and gentle outer slopes, it has been interpreted as the end of an anticline pitching eastward.[6] From this point to Easton the route traverses the Highland belt. Like much of the Highland area this section consists of pronounced ridges and valleys. The valleys in most cases are broad and flat-floored, and fertile because of their limestone soil. Most of the valleys are synclinal in structure,[7] either complete or broken by faulting along their sides. In this way the Paleozoic limestones and shales have been carried into the crystalline mass and are now preserved below the level of erosion. The first and most pronounced of the Highland ridges is Musconetcong Mountain.

[4] The Triassic Lowland, in the region of the Watchung Ridges, is a shallow basin in structure, and is shaped something like a flat gravy bowl.

The rocks dip from all directions toward the center of the basin. The steep side of the ridge therefore faces outward.

[5] A "fault line" scarp is an escarpment due to erosion along a previously formed "fault" or displacement.

At first there is an original fault scarp. This disappears if the region is peneplaned. Then erosion of the weaker beds on one side of the fault forms a "fault line" scarp.

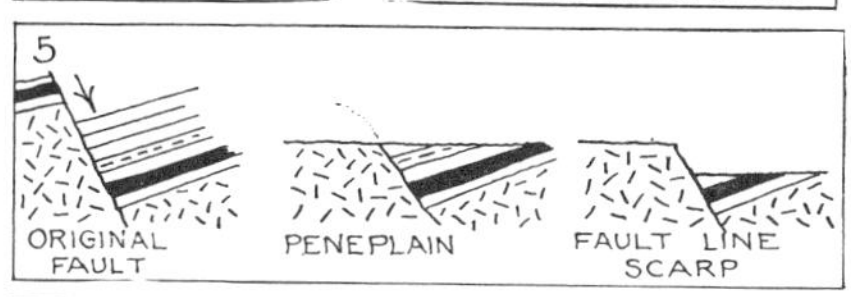

[6] An anticline is an up-arched structure or fold. A pitching anticline is an arch or fold which slopes downward at one end.

Erosion of a pitching anticline results in a curved ridge, the steep side of which faces inward toward the axis of the anticline.

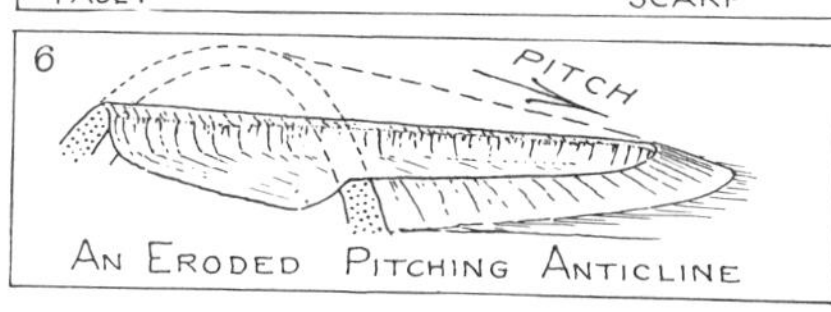

[7] A syncline is a trough-shaped structure or downward fold.

This is just the opposite of an anticline or arch. The basin shown in Note 4 above is also a syncline. See Note 11 on page 30 and Note 12 on page 31 for other examples.

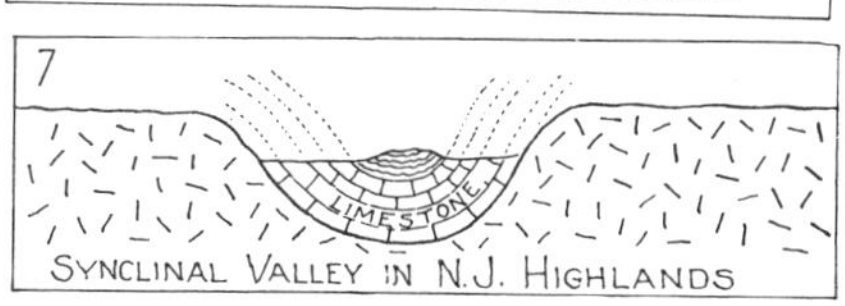

Like the other ridges it is transected in several places by narrow but important gaps through which railroads and roads run, and over which the air route lies. The trip over the Highland hills will probably be slightly bumpy, and then soon the Delaware River appears beneath. Just above Easton the rapids in the stream betoken the presence of hard rock ledges. Some actual outcrops may be seen conforming in strike with the Highland features.

The observer now beholds the broad expanse of the Great Valley, bounded on the west by Kittatinny Mountain. About 20 miles to the north the gap cut by the Delaware River across this ridge can easily be seen. (Fig. 14)

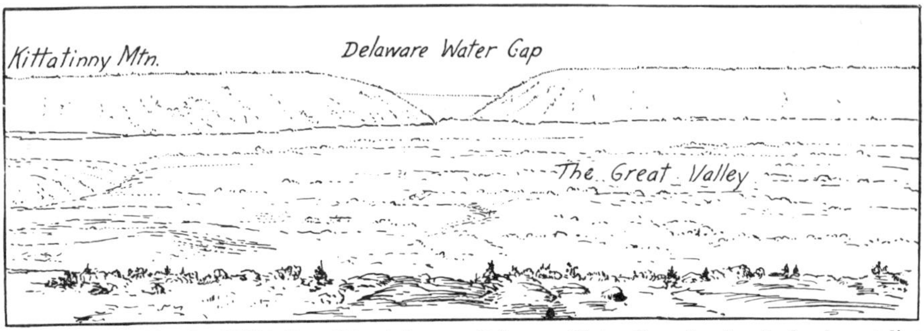

FIG. 14. Sketch of Kittatinny Mountain near Delaware Water Gap, showing its level crest line, as viewed from the east over the Great Valley.

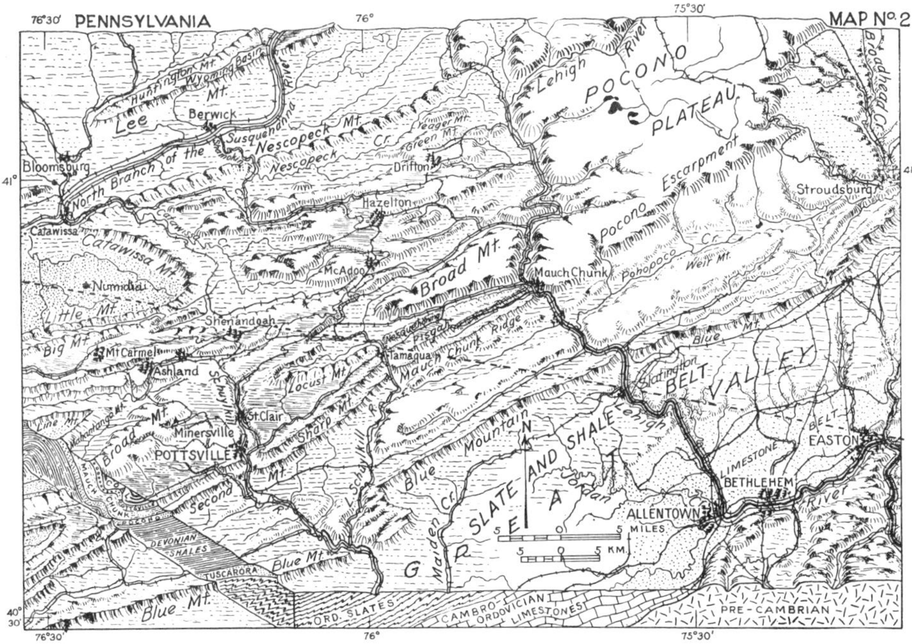

Route Map No. 2

EASTON TO SHAMOKIN

The Great Valley. At Easton the air route leaves the New Jersey Highlands, with their crystalline rocks, and enters the Folded Appalachians. The Great Valley, some 15 miles in width, lies directly ahead. Beyond, on its western side, is Kittatinny or Blue Mountain, winding like a gigantic snake across the country. The Great Valley, here called the Lebanon Valley, presents two slightly contrasting zones, an eastern and a western. The eastern zone is a region of limestone, the lowest of the Paleozoic beds of this region, usually termed the Cambro-Ordovician limestones. The western belt is underlain by slate and shale, and stands somewhat higher than the limestone area. It is more dissected and has a more rolling aspect than the comparatively flat limestone section, and its buff yellow soils contrast with the reddish residual limestone soils observed immediately after passing Easton. A view to the south reveals the cities of Allentown and Bethlehem, situated on the Lehigh River where that stream impinges against the crystalline hills of the Reading Prong before turning north to join the Delaware.

The Lehigh Valley is one of the most important routes from the anthracite coal region to the Atlantic seaboard. It gives name to the Lehigh Valley Railroad, which follows the course of the Lehigh River from Easton westward. Looking down upon this stream, black with coal dust, the air traveller is almost certain to see one or more long trains of coal on the tracks which parallel each bank.

Many slate quarries may be noted in the western half of the Great Valley, the largest being at Slatington (Fig. 15), making this the most important slate producing section of the United States, and giving Pennsylvania the leading rank in that commodity. From the vicinity of Slatington, where the Lehigh River is crossed, there is to be had a magnificent panorama over Blue Mountain, which stands out very prominently and runs as far to the north and to the south as the eye can see. Its crest and slopes

FIG. 15. The Great Valley at Slatington, Pennsylvania. In the foreground are some of the old slate quarries, now filled with water.

are heavily forested, except for some bare patches that look like sand, but which in reality are vast talus[8] slopes and masses of broken conglomerate.

The Anthracite Coal Basins. From Blue Mountain to Shamokin (Map 3) the air route traverses several of the anthracite coal basins.[9] The anthracite district lies east of the Susquehanna River. It consists of several synclines which trend northeast-southwest and fork out at the two ends into a number of prongs. The Pottsville Basin is the largest and easternmost of the individual synclines which make up this synclinorium. It branches toward the southwest, one fork going on either side of the Peters Mountain anticline (shown best in the southeastern corner of Map 3). Toward the northeast a long prong extends from the Pottsville Basin to Mauch Chunk. In the axis of this prong lies the city of Tamaqua, almost directly under the air route. Two resistant formations outcrop in pronounced ridges around the margins of the Pottsville basin. The uppermost (that is stratigraphically) is the *Pottsville conglomerate,* which forms Sharp Mountain, Pisgah Mountain, Nesquehoning Mountain,

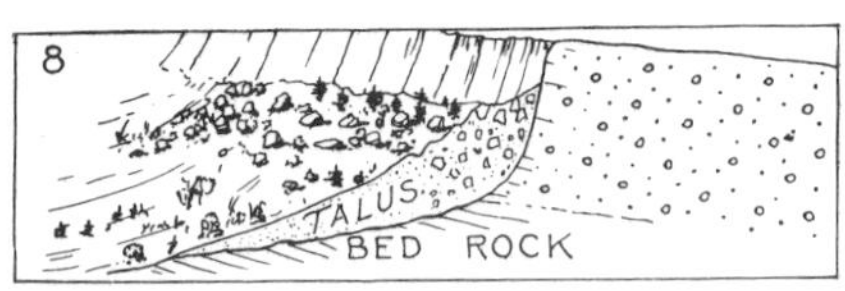

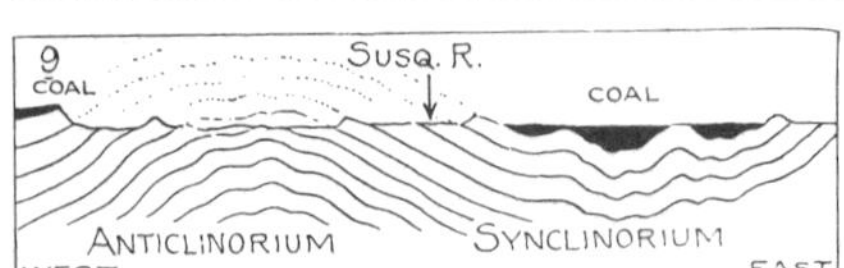

8 Talus is the accumulation of loose rock at the foot of steep cliffs. Near the cliff it consists of large angular blocks which have broken off along the joint planes. Farther away the talus is made up of smaller fragments which grade into fine alluvium carried there by running water.

9 In no part of the world are folded rocks on a large scale so magnificently displayed as they are in the State of Pennsylvania. For over a century this has been classic ground for the geologist. The variety and intricacy of these forms as viewed from the flying plane are displayed in the most striking way. Familiarity with the region is almost necessary in order to grasp the significance of all that may be seen during the fleeting moments of time available on an air trip. A fair understanding of the structure and topography of eastern Pennsylvania is therefore desirable. In a general way it may be said that the eastern half of the Pennsylvania folded belt is a synclinorium or a trough with wrinkles, the western half, an anticlinorium or arch with wrinkles. Erosion has removed the coal from the crest of the anticlinorium but it still remains in the deeper parts of the synclinorium. Refer also to the Figures in the supplementary chapter, especially Figs. 91, 92 and 93.

FIG. 16. Shamokin and other coal-mining towns in the anthracite region of eastern Pennsylvania. The parallel ridges of the Folded Appalachians are visible in the distance.

and several others. Above this conglomerate are the non-resistant coal measures which outcrop continuously from near Mauch Chunk southwestward to beyond the limits of Map 2. Several coal-mining[10] towns lie along this belt (notably Tamaqua, St. Clair, Minersville and Pottsville), joined by railroads and roads. It is indeed a veritable hive of activity.

FIG. 17. The ridges and valleys of eastern Pennsylvania, near Shamokin. One of the ridges is transected by a water gap.

[10] Anthracite or hard coal occurs in the folded mountains of eastern Pennsylvania, whereas bituminous or soft coal occurs in the western or plateau area of Pennsylvania, where the rocks have not been disturbed. Folding and pressure changes soft coal into hard coal.

FIG. 18. The ridge and valley section of eastern Pennsylvania. View looking northeast from near Pottsville, showing some of the linear ridges, transected by water gaps.

The lower of the two resistant formations involved in this basin is the *Pocono sandstone* which forms Second Mountain and Mauch Chunk Ridge. North of the Lehigh Gap at Mauch Chunk the synclinal structure comes to an end, and the Pocono sandstone flattens out to form the Pocono plateau, the eastern margin of which is called the Pocono Escarpment.

West of Mauch Chunk is Broad Mountain, which constitutes the flat top of a southward-pitching anticline of the Pocono formation.[11]

West of the Pottsville Basin and separated from it by a slightly upwarped area is another long syncline, the Shamokin Basin, in which lie the towns of Shenandoah and Mt. Carmel. It comes to an end a few miles west of Shamokin (Map 3) where

FIG. 19. The ridges and valleys of eastern Pennsylvania, a view looking south near Shamokin.

[11] A geological section through this region makes this clear.

The Pottsville conglomerate and the Pocono sandstone are resistant and form the ridges. The intervening Mauch Chunk shale is weak and forms the valleys. Profound erosion has removed all of the upper beds from the top of the anticline.

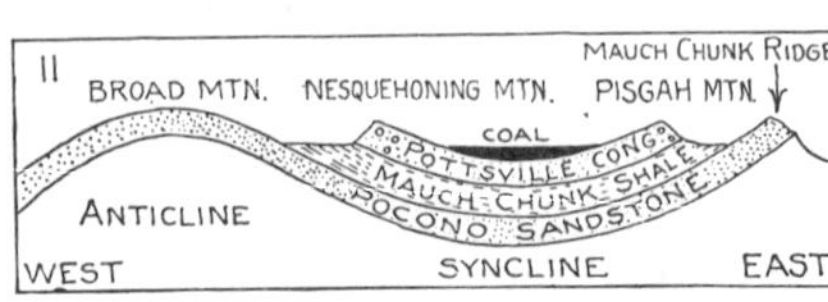

Big Mountain and Mahanoy Mountain unite to form the nose of the eastward-pitching syncline. Line Mountain and Little Mountain (also on Map 3) unite in a similar manner. The eastern end of the Shamokin Basin opens out northward and forms the broad Hazelton Basin. These two basins, the Shamokin and the Hazelton, constitute essentially one large structural feature. Each one contains many coal-mining cities, with a congestion of railroads and roads. Unlike the Pottsville Basin, which is a comparatively simple syncline, the Shamokin Basin is made up of numerous minor folds. These add several complications to the details of the topography. Some of these details can be observed from the air, especially if the plane flies along the axis of the main fold over Shamokin (Fig. 16) and the other coal-mining towns, instead of along the marked route several miles to the north. In some places the outcrops of certain beds can be followed by the rows of clay pits which have been excavated in them.

Views to the north from the region east of Shamokin reveal clearly the southern end of the Wyoming (or Scranton) Basin, where Lee Mountain and Huntington Mountain unite to form the nose of the northeastward-pitching syncline with its distinct canoe-shape. This structure is transected by two magnificent gaps cut by the Susquehanna River, only one of which appears on Map 2. In this basin just beyond the map, but almost within sight is the city of Wilkes-Barre. Views southward from this general region reveal many of the zigzag features represented on Maps 2 and 3.[12] The different ridges succeed each other in rapid succession (Fig. 17). There are in fact eleven major ridges, all very prominent topographic features, lying between the line of flight and the Great Valley, 25 miles to the south (Figures 18 and 19). These may all be counted on the maps, and many of them identified from the plane.

12 The erosion of pitching anticlines and synclines produces the remarkable zig-zag mountains of central Pennsylvania, called by the Indians, the "Endless Mountains."

In the accompanying sketch both the anticline and the syncline pitch toward the right. Only one resistant layer is shown. In the Pennsylvania ridges, however, there are several hard layers and these form, therefore, a similar number of parallel zig-zags.

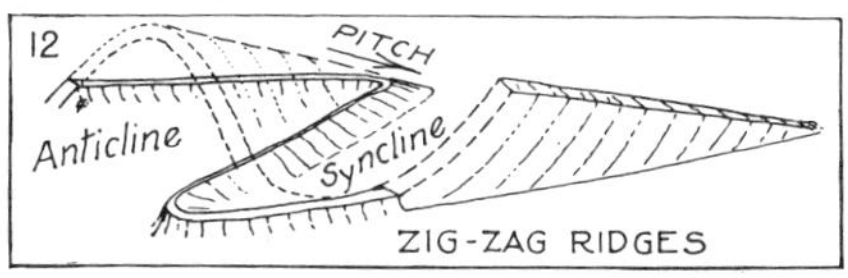

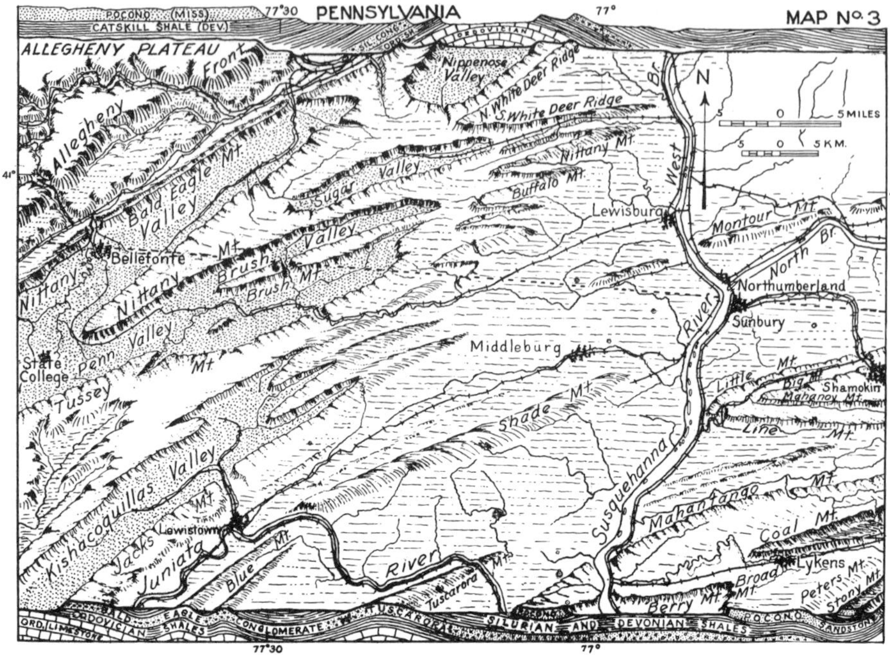

Route Map No. 3

SHAMOKIN TO BELLEFONTE

Zigzag Ridges. The chief features of interest immediately after passing Shamokin are the zigzag ridges (see Note 12 on page 31) lying to the south of the air route. Line Mountain and Little Mountain converge and come to an abrupt end at the Susquehanna River. These are formed of the Pocono sandstone. The next higher resistant formation, the Pottsville conglomerate, in a similar manner determines two converging ridges, Mahanoy and Big Mountains. Still other converging ridges, much farther south, can also be seen.

The Susquehanna River. At Sunbury the route crosses the Susquehanna River (Fig. 20) where the West Branch and the North Branch unite. The West Branch drains the plateau region lying to the west. The North Branch and its tributaries drain the anthracite coal basins. The water of the West Branch is brown and muddy in color, that of the North Branch is black. As far as the eye can see down the river the two currents remain separate. In fact, even at Harrisburg, where the fine coal dust is dredged from the river bed, over 50 miles downstream, this condition is still true.

The Eroded Anticlines West of Sunbury. The belt of the Folded Appalachians lying between Sunbury on the Susquehanna River and the Allegheny Front, just west of Bellefonte, is an anticlinorium, whereas the region east of the Susquehanna is a synclinorium. The western belt, too, has a simpler aspect than the eastern belt because only one resistant formation is represented. This is the Bald Eagle conglomerate which corresponds with the Tuscarora formation farther east. West of the Susquehanna, the Pocono and the Pottsville beds have been removed although they occur again in the plateau which lies beyond (see Note 9 on page 28). For the

FIG. 20. The Susquehanna River below Sunbury, looking southeast. Beyond the river, on its eastern side, are many of the wooded ridges of the Folded Appalachians.

first 20 miles west of Sunbury the terrain is distinctly flatter (Fig. 21) than that encountered elsewhere in the folded belt. This broad zone is developed upon Silurian and Devonian shales of low resistance. The Silurian rocks represented by the Clinton beds, are red, due to their origin as desert deposits. Contrasting with these deep red soils are the buff and orange soils of the limestone valleys a few miles to the west.

The eastern part of the up-arched area west of the Susquehanna is marked by a splendid series of anticlines pitching toward the northeast (see Fig. 93). Half a dozen or more are of very striking character with their rounded surfaces, and their

FIG. 21. The rolling topography west of Sunbury, a region of weak shales in which no pronounced ridges occur.

FIG. 22. An eroded pitching anticline. This anticline is pitching toward the left. The two ridges, formed by the limbs of the anticline, converge and join to form a single ridge.

long tapering ends plunging beneath the plains. The crests of the arches have in almost all cases been eroded, so that valleys occur along the axes (Fig. 22. Also Note 6 on page 25). Brush Valley and Penn Valley as well as Nittany Valley in which Bellefonte lies (Fig. 23), all crossed by the air route, are of this type, and there are several others also shown on Map 3. The steep inner slopes and the gentle outer slopes of the eroded pitching anticlines offer a striking contrast with the steep outer slopes and more gentle inner slopes of the pitching synclines, already noted farther east. The synclinal axis between Penn Valley and Brush Valley is occupied by Brush Mountain, a synclinal mountain which pitches to the northeast. The nose of this pitching syncline, lying just to the south of the air route, is in full view.

FIG. 23. Nittany valley at Bellefonte airport. The Lower Paleozoic limestones have been eroded along the axis of a broad anticline and now are exposed in the floor of the valley. The ridge in the distance, forming the eastern rim of the valley is Nittany Mountain. It is made up of eastward-dipping conglomerates.

North of Brush Valley, and 5 to 10 miles north of the air route, is Sugar Valley, an almost completely enclosed basin, developed on the axis of an anticline. Still farther north is Nippenose Valley, of similar origin, but this is beyond the range of easy vision.

Between Brush Valley and Nittany Valley is a synclinal structure forming Nittany Mountain which comes to an abrupt end 8 miles south of Bellefonte, not far from Pennsylvania State College. Nittany Mountain, rising boldly on the east side of Nittany Valley (Fig. 23) was, in the early days of air mail, often a dreaded point for pilots.

All of the ridges in central Pennsylvania are wild and heavily timbered but the intervening rich limestone valleys are extensively cultivated and offer a charming prospect when viewed from aloft. In one or two places limestone quarries may be seen, and near Bellefonte there is a large sink,[13] almost a mile across, its margins inward-sloping with their orange soils intricately gullied and bare of vegetation.

[13] "Sinks" or "sink holes" are common in limestone regions. Underground drainage, due to the solubility and porosity of the limestone, permits the surface water to enter subterranean channels through depressions on the surface which gradually enlarge and extend over many acres.

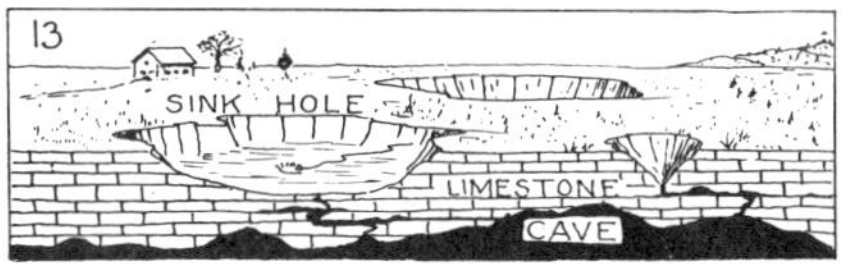

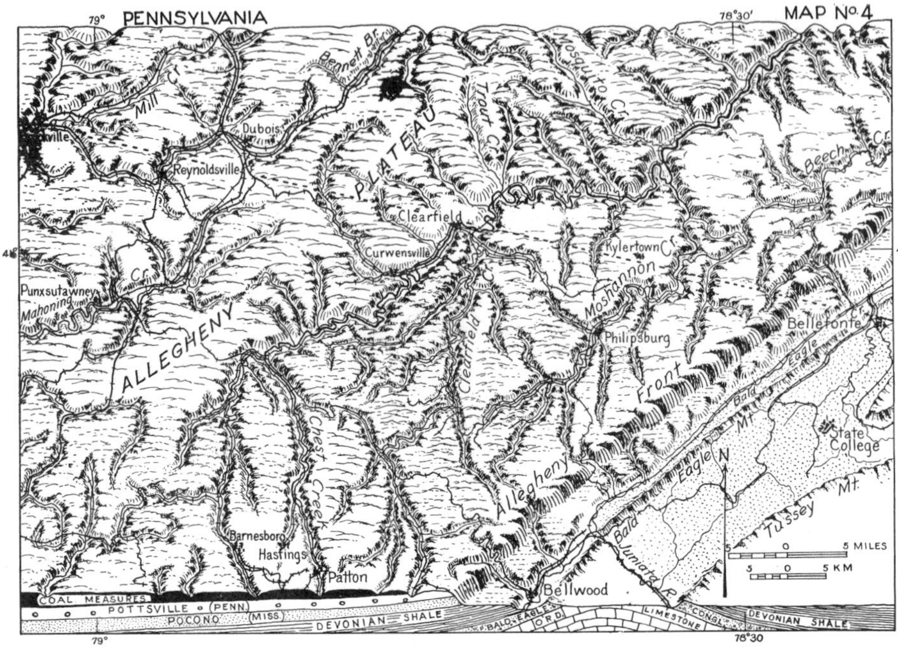

Route Map No. 4

BELLEFONTE TO BROOKVILLE

Bald Eagle Ridge. Immediately west of Bellefonte and visible from the airport is the last one of the folded ridges, Bald Eagle Mountain. The formation which accounts for this ridge is known as the Bald Eagle conglomerate. It dips strongly toward the west beneath the plateau series. Although the east and west sides of this ridge have approximately the same steepness of slope, the two sides differ in appearance. The western side is continuous and uninterrupted, this being the dip slope. On the eastern side there is a series of bench-like spurs which results from the dissection of a secondary resistant member underlying the main ridge-former.[14] Bald Eagle Creek occupies the subsequent[15] lowland between Bald Eagle Mountain and the Allegheny Plateau.

The Allegheny Plateau. The imposing Allegheny Front now commands attention. The escarpment itself, formed largely by the Pocono sandstone (Mississippian or Lower Carboniferous) rises almost vertically 800 feet above the platform of Catskill shales at its base. Both the plateau upland and the foothills or "foreknobs" are maturely dissected and heavily forested (Fig. 24). The platform of Catskill shales is much finer textured[16] than the body of the plateau with its coarser features resulting from the erosion of the massive Pocono beds. Notches in the plateau front are not numerous, nor do they cut back far into the upland. Nevertheless it is by these notches that the railroads ascend to the country beyond. Just north of the air route a branch

[14] The detailed aspect of Bald Eagle Mountain is suggested in the accompaning sketch.

[15] A "subsequent" lowland or valley is one developed along a belt of weak rock.

[16] Fine textured topography means many streams and small details. Coarse textured means fewer but larger streams in a given area.

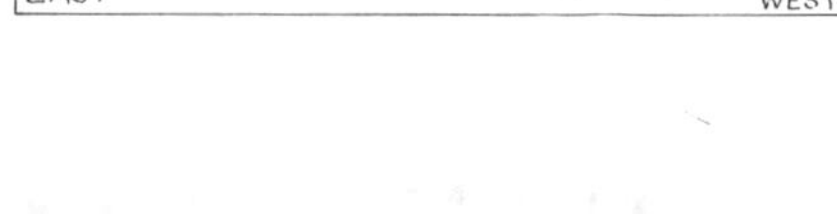

FIG. 24. The forest-clad Allegheny Plateau and the Allegheny River.

of the Pennsylvania Railroad winds up through one of the more open valleys which dissects the eastern face of the plateau and, crossing the divide, gains entrance to the valley of Moshannon Creek. The belt of country between this stream and the front of the plateau is wild and deserted. From aloft it looks like a vast carpet of green except for the brown and barren patches where forest fires have destroyed the timber. The regular branching of the dendritic drainage pattern gives this region a uniformity of appearance unlike that farther east where the plan of the streams is controlled by the folded structure.[17] From the air, too, the observer has clear evidence of the almost horizontal attitude of the plateau beds in the contour-like pattern of the minor scarps and lines of vegetation. In fact parts of the upland seem to be capped with mesas.[18]

The top of the plateau is covered by the Pottsville conglomerate, separated from the underlying Pocono sandstone by the weak and comparatively thin Mauch Chunk shale. It is the variation in the character of the conglomerate beds which causes the development of the terraces on the divides and upland surfaces. In general the vegetation along the valley walls is a much deeper and richer green than on top of the plateau, due to the more abundant moisture available there.

The valleys of the Moshannon and the Susquehanna are fairly open and show considerable human activity in the form of agriculture and many towns and villages.

[17] The streams in the folded belt have a trellis pattern, because many of them follow the belts of weaker rocks.

These weaker belts run parallel with each other and are separated by long ridges across which the transverse streams pass through water gaps. "Trellis" means like a vine. "Dendritic" means like a tree.

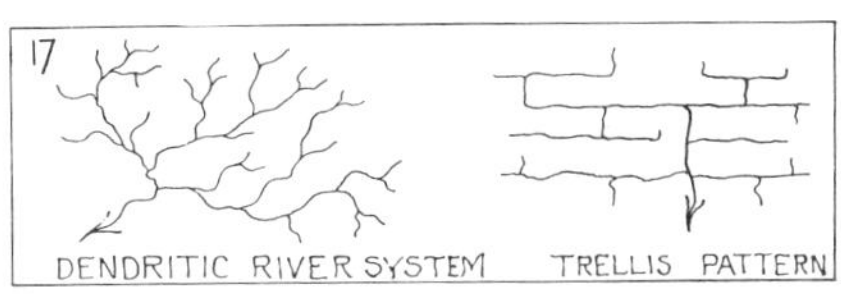

[18] Although "mesas" are typical of the western states, they may develop wherever formations lie horizontally. The vegetation and soil cover in the eastern United States usually conceals their sharp outlines.

Evergreen trees usually grow on the steeper slopes and hardwood or deciduous trees on the intervening flatter spaces.

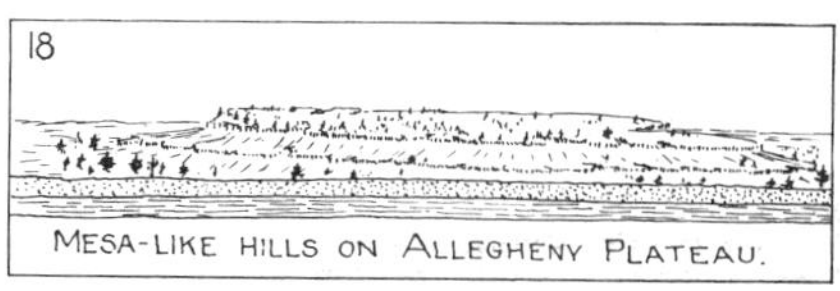

Some of the streams, notably the Moshannon, carry a flood of rusty water derived from the iron-bearing beds in the plateau rocks.

The Susquehanna River, which is crossed at Clearfield, presents a fine display of incised meanders although there is less difference between the undercut and the slip-off[19] slopes of the meander spurs, than may be noted in the case of several of the smaller streams in this region.

Bituminous Coal Mines. Owing to the broad synclinal structure of the entire Allegheny Plateau, the youngest formations, which are the coal horizons of the upper Carboniferous, are found in its central part.[20] The Plateau syncline also pitches or slopes to the south, with the result that the whole plateau has the structure of an elongated basin. The center and therefore the deepest part of this basin is in south-western Pennsylvania and West Virginia. Inasmuch as the air route crosses the northern part of the plateau region it does not encounter the important coal fields. However in a few of the valleys along the route coal mining activities may be noted, as for example near Reynoldsville and DuBois. Little spur railroads penetrate the various side ravines and gulches where the coal beds outcrop as horizontal bands on the valley sides. Much of this coal is also shipped by truck direct to the consumer.

[19] The undercut side of the spur faces up the valley; the slip-off slope faces down the valley. This contrast is clearly shown by several of the streams in the Allegheny Plateau crossed by the air route.

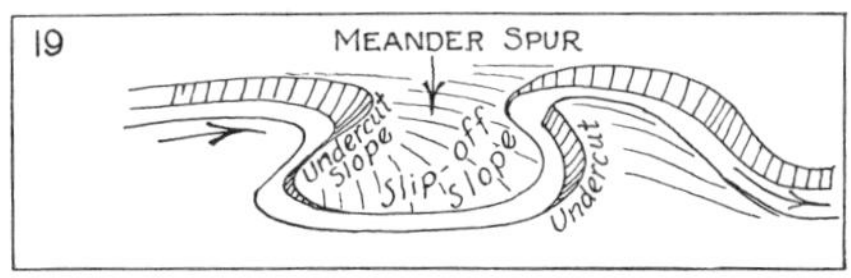

[20] Practically all the coal fields of the eastern United States, such as the western Kentucky field and the Michigan basin, are synclines or shallow troughs in which the coal has been preserved from erosion.

Between these coal fields are up-arched areas from which the coal has been eroded.

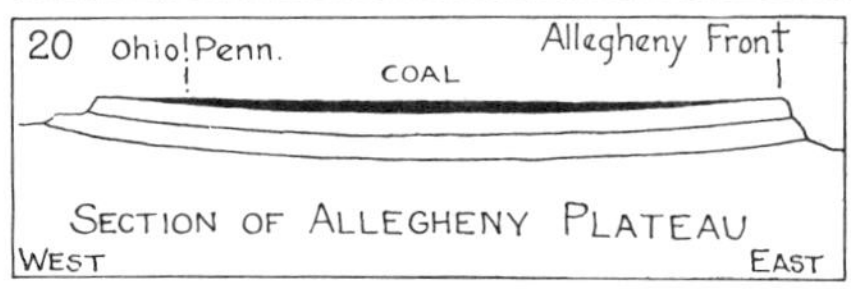

Route Map No. 5

This map shows the location of the numbered light beacons marking the airway across the plateau. These beacons, situated about ten miles apart, do not form a straight line, as they have to be placed in advantageous positions. Plate VII shows the beacons for the entire airway.

BROOKVILLE TO MERCER AND SHARON

The Rugged Allegheny Plateau. West of Brookville (Map No. 5) the route passes over one of the most rugged sections of western Pennsylvania. The streams in this part of the plateau have deeply entrenched meanders. The narrow valleys are not habitable and most of the activities are on the plateau surface. Clarion, for example, is thus situated, overlooking the gorge of the Clarion River. A landing field, just east of the city, is on a flat part of the plateau which drops 400 feet almost vertically to the river below. The Clarion River valley is too narrow and canyon-like and without sufficient valley floor to serve as the route for any railroad or even any public road. The Allegheny, with its occasional patches of alluvial bottom, is used by one of the divisions of the Pennsylvania Railroad, which follows its eastern bank.

Clarion is a small city of less than 4,000 people. The tower of the Clarion County court house situated there is occasionally used by pilots to check their altimeters, as the top of the steeple is just 1,500 feet above sea-level. This city, too, boasts the largest milk bottle plant in the world, an example of the heavy glass-making industry which thrives in this region, where quartz sand, coal, oil, and gas are all available.

The Oil Fields of Pennsylvania. Several of the oil fields of western Pennsylvania are crossed during this part of the trip. They are not situated with any respect to the valleys which dissect the plateau. The valleys are unrelated to the minor warping and the subdued folds of the region, which trend in a northeast-southwest direction. The streams have apparently incised their courses below an old peneplain level which bevels the different formations. The oil fields, on the other hand, follow the axes of the anticlines or at least the general trend of the structure.[21] One of these axes runs in a northeast-southwest direction through the city of Clarion. Another, and much larger one, runs through Knox, and a third through Lamartine. These oil fields are crossed in rapid succession by the air route. There are literally hundreds of oil wells in these so-called pools. In one of the pools southwest of Knox, over which the air route passes, there are 40 to 50 wells to the square mile. Unfortunately these wells can hardly be seen from the airplane except at low elevations. Only some of the wells have derricks or structures above them and these are usually in a dilapidated condition. The site of each well is marked by a simple pump which does not stand more than a few feet above the ground. The pumps are operated by levers which run from a central power station to the different wells. Sometimes they are run by compressed air, but little of these devices can be noted by the speeding observer.

To the air traveller the location of farmhouses in this region may seem to bear no relation to topography or geology. The majority of the farmhouses, however, are located very near a coal outcrop, not for convenient access to the coal, but because along coal outcrops there are many springs, and springs are often determining factors in locating farmhouses. Coal beds are commonly underlain by clay, and this forms an impervious layer along which the ground water travels until it reaches the surface of the ground at the side of a hill.

The Glaciated Part of the Plateau. The Allegheny River is crossed at Kinnerdell, where one of the intermediate landing fields may easily be seen on the top of a northward-projecting meander spur. About 8 or 10 miles west of this point the route passes over the terminal moraine which here runs north and south. A few miles southeast of Stoneboro (see Stoneboro Quadrangle of the U. S. Geological Survey), where the route crosses the moraine,[22] it consists of a belt of small hills two to three miles wide. Numerous small elevations and depressions mark its course. It is little if any more covered with trees than the adjacent country. Three or four ponds occupying kettle holes lie almost directly under the course of the airplane, but altogether the features of the moraine are rather insignificant and apt to be lost in the bolder topography of the region. However, once the moraine is crossed, swamps appear. These are usually covered with a growth of trees and are not always to be recognized as swamps. South of Greenville (see Route Map No. 6 and also Shenango Quadrangle of the U. S. Geological Survey), two of the old glacial channels may be noted, now occupied by the Shenango River and its tributary, Pymatuning Creek. These are both misfit[23] streams flowing in a wide flat-floored valley, except, just west of Fredonia, where the valley has been blocked with glacial drift.

21 Oil often accumulates at the crest of an anticline because, being lighter than the ground water in the rocks, it rises until it escapes at the surface or is held in by an impervious formation.

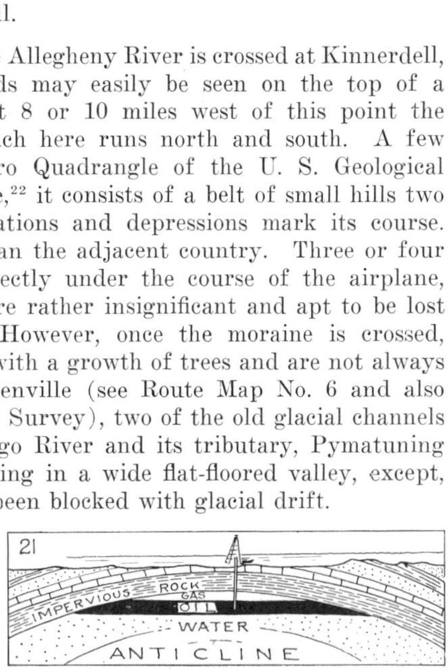

22 A moraine is a belt of low hills made up of soil and stones deposited by the ice sheet at the end of glacial time. Hollows or depressions often containing small ponds are known as "kettle holes."

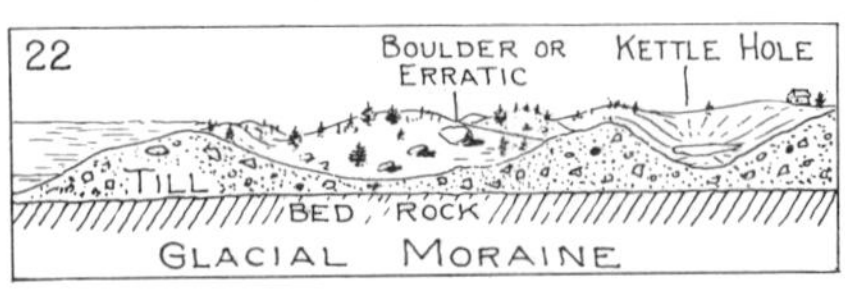

23 A misfit stream is one with many small meanders flowing in a valley originally cut by a much larger stream. This is due in many cases to the change in the behavior of a river which carried a large volume of water during glacial time and has since dwindled to a mere brook.

For several miles along this part of the course the golden dome of the Mercer County court house at Mercer may be seen gleaming in the sun, six miles or so south of the air route.

The agricultural activities of the region between Brookville and Mercer are very limited. In fact, this plateau portion of the eastern United States is noteworthy because of its dearth of products. Buckwheat and oats, two crops which do well on rough topography and stony soil, are represented here. Haying is important and the shipping of eggs constitutes a source of cash income.

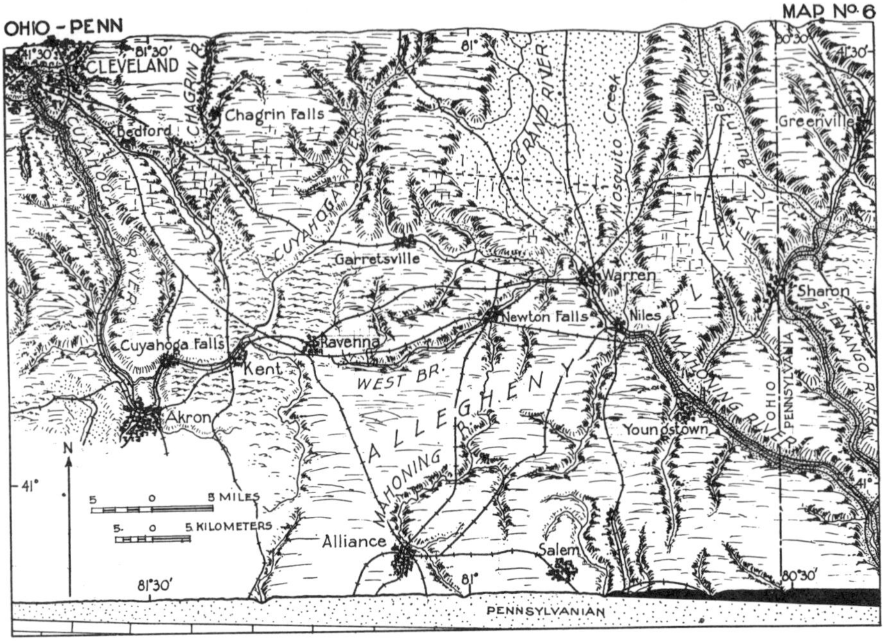

Route Map No. 6

SHARON TO CLEVELAND

The Township System of the Middle West. A few miles west of Greenville the route crosses the state boundary line into Ohio. Ordinarily the passage from one state to another is not marked in any way, but in this case the air traveller can tell the moment he is over the line. On the Pennsylvania side the roads conform with the topography by following the valleys and the ridge tops. In Ohio the roads follow the checkerboard or rectangular system established in 1785 when the so-called Northwest Territory was surveyed. A few facts about this system will help the traveller to appreciate the marvelously regular layout of roads and farms which he now enters upon and which spread westward, more than a thousand miles to the Rocky Mountains. The essential idea was to divide the western United States into a huge gridiron. This was done by establishing a number of principal meridians (Fig. 25) at various intervals, one being along the western Ohio boundary (the First Principal Meridian), another through central Indiana (the Second Principal Meridian), a third through Illinois (the Third Principal Meridian), and others through Iowa, Nebraska, and other states. East and west of the principal meridians, north and south rows of townships, called ranges, were laid off. Each principal meridian, together with the system of townships based upon it, is independent of every other principal meridian, and where two systems come together irregularities are found.

Correction Lines. A base line is run at right angles to each principal meridian. Tiers of townships, each 6 miles square, are laid off north and south of these base lines (Fig. 26). A township may therefore be designated, for example, as Tp. No. 2 South, Range 3 West of the Third Principal Meridian. The eastern and western boundaries of townships are, as nearly as may be, true meridians, and when they have been extended northward through several tiers, their convergence becomes

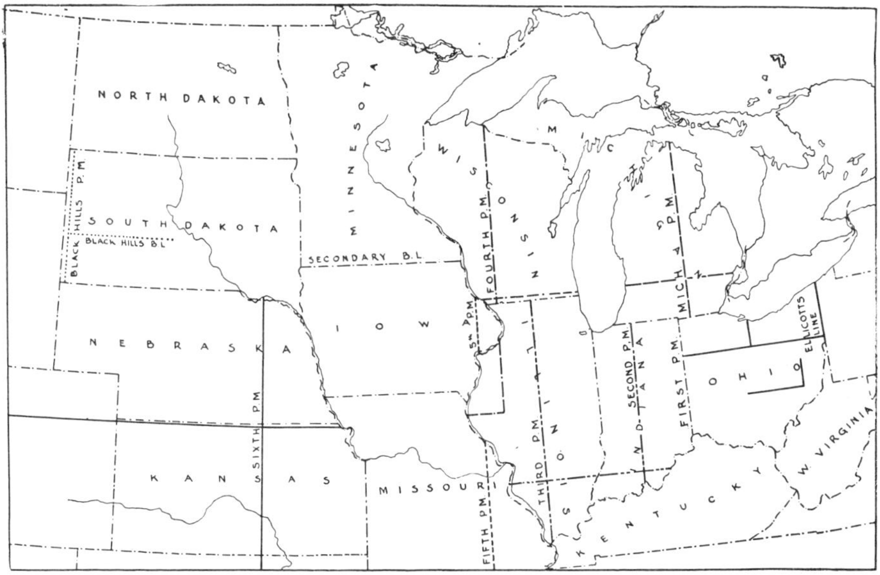

FIG. 25. Map showing the Principal Meridians (marked P.M.) and the Base Lines (marked B.L.) corresponding with them, used as a basis for laying off the townships in the region lying between the Appalachian Mountains and the Rockies.

considerable. To prevent this diminution in size of townships to the north of the base line, about every four tiers a new parallel is run, along which 6-mile measurements are made for a new set of townships. These are known as correction lines.

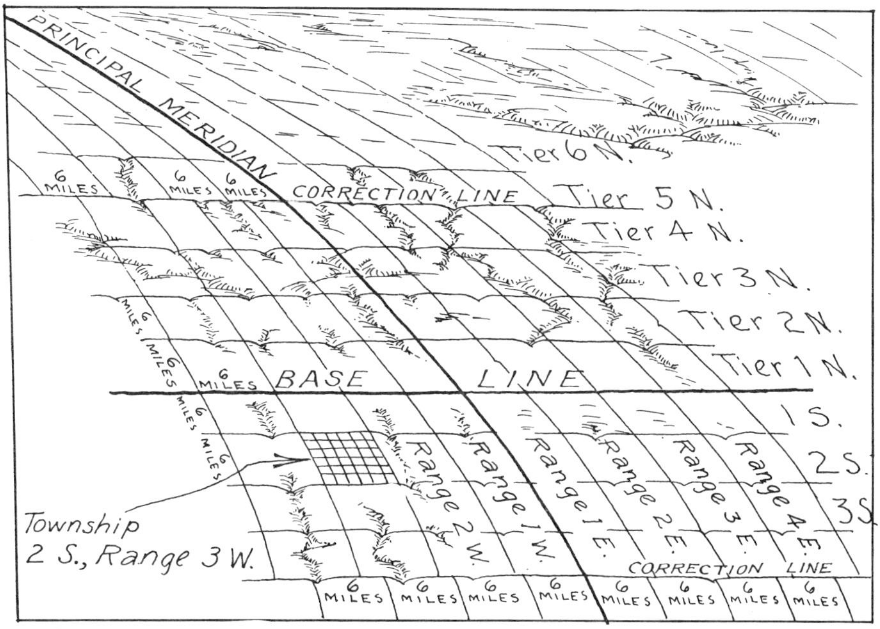

FIG. 26. Diagram showing "tiers" and "ranges" of townships arranged with respect to a "Principal Meridian" and "Base Line." Two "correction lines" are shown, four tiers north and four tiers south of the base line.

Since public roads are usually built on section and quarter section lines, wherever a north-south road crosses a correction line there is a "jog" in the road, as the accompanying figure will show (Fig. 26). To the traveller in the air these jogs are very much in evidence because many of them can be seen at the same time whereas the man on the road below knows only of the single jog which he is making and does not always realize that it is one of a series.

Size of Farms. Each township consists of 36 square miles, since the township is 6 miles on a side. Each square mile or section consists of 640 acres. A quarter section of 160 acres is one-half mile on a side. This in turn may be subdivided into quarters of 40 acres each. Many farms in the middle west comprise 40 acres or multiples thereof.

In eastern Ohio the grid-like arrangement is not quite so regular as will be found farther west in Indiana and Illinois where the land is more nearly level. Nevertheless the plan of the roads which in general follow the section lines is distinctly rectangular.

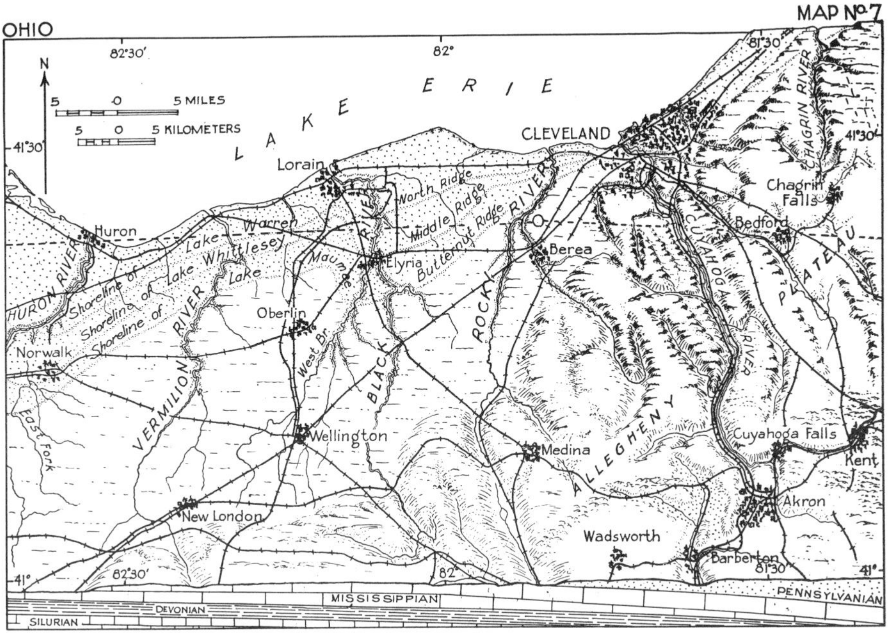

Route Map No. 7

THE CLEVELAND REGION

The Rolling Plateau of Eastern Ohio. The Allegheny escarpment (this is not the same as the Allegheny Front) which runs east and west across central New York State swings southward along the southern shore of Lake Erie and then almost directly south across Ohio. Unlike the New York portion, the Ohio part is not a pronounced scarp but is rather a zone of rolling hills, by which the plateau surface descends gradually toward the north and toward the west. East of Cleveland two or three fairly large northward-trending valleys have been opened out below the plateau surface. The valley of Grand River (Map No. 6) and that of Chagrin River are the most conspicuous. Their north-south trend and the general parallelism of the streams in this part of the state immediately strikes the eye, because it is in contrast with the irregular dendritic (See Note 17 on page 37) branching of the streams farther east in Pennsylvania. This parallelism is due to the fact that along the western margin of the Plateau the beds dip slightly toward the east. The north-south strike of the rocks apparently affects stream position. These streams, as well as the Cuyahoga at Cleveland, and other smaller streams west of Cleveland, have cut entirely through the Mississippian and Pennsylvanian beds, and in most cases are cutting gorges into the black shales of Devonian age. The observer in the air as he passes over the various stream valleys upon entering and leaving Cleveland can not fail to notice these dark colored formations and the horizontal attitude of their bedding planes.

Cleveland. Cleveland itself stands upon the delta of the Cuyahoga River which was built out during glacial time into Lake Maumee. This great lake preceded Lake Erie and covered somewhat more territory. With a population of almost 1,000,000 Cleveland is one of America's largest cities. The air route lies too far to the south for the traveller to see many of the important landmarks, but on a clear day he may descry the great Terminal Tower which rises over 700 feet toward the sky. Beyond lies the broad expanse of Lake Erie where Cleveland secures its water through a tunnel which reaches five miles out beneath the lake.

The railroads entering Cleveland from the east follow either the shore of Lake Erie, utilizing the flat beach of Lake Maumee, or else come down the Cuyahoga Valley and its tributaries. These latter lines, namely the "Erie," the "Wheeling and Lake Erie," the "Pennsylvania," and the "Baltimore and Ohio," are crossed by the air route. They connect the important lake port of Cleveland with the industrial centers and coal fields of eastern Ohio and western Pennsylvania. Mention may be made especially of the large volume of iron ore which comes into this region from the head of Lake Superior (Fig. 27). As an adjunct to rail transportation the Ohio

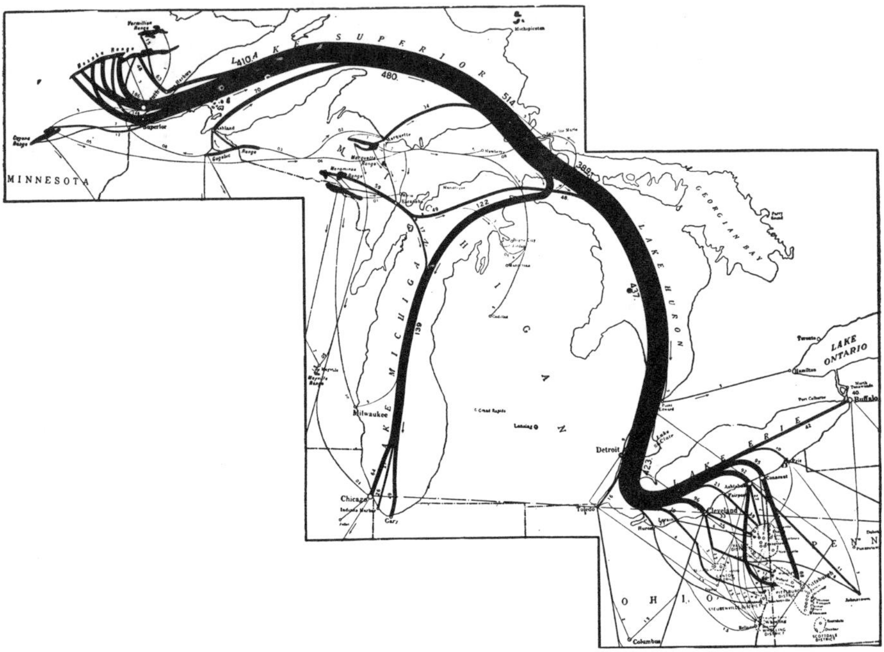

FIG. 27. Map showing the movement of iron ore from the mines near Lake Superior to the furnaces near Chicago and in the Lake Erie region.

Canal runs along the flat valley floor of the Cuyahoga River on its eastern side to join the Ohio 100 miles to the south. This is used for the movement of heavy commodities such as coal and ore between the Great Lakes region and the Ohio Valley.

West of Cleveland also the railroads and roads converge upon the city but, owing to the flat character of the land, they are much less influenced by the topography.

Ancient Beach Ridges. The main highways between Elyria and Cleveland are situated upon the old beach ridges of Lake Maumee. Three of these roads can easily be seen from the air, although the ridges themselves are too obscure to be recognized. (See Berea and Oberlin sheets of the U. S. Geol. Survey.) Between the ridges the land is low, flat and marshy, and is usually undeveloped, being left in natural meadows or covered with trees. Sheep are grazed in these localities. On the tops and slopes of the ridges, for the width of half a mile or so, the land is cultivated, the farmhouses being all situated along the main road. The difference in level between the ridge tops and the lower land on either side is about 30 feet.

Immediately west of the Cleveland Airport is the gorge of the Rocky River. In this gorge, directly under the air line may be seen two mesa-like hills, formed as the result of two successive captures or cut-offs.[24] The rectangular and straight-sided form of the valley walls is due to the very regular jointing of the black shales into which Rocky Creek is cutting.

[24] These features are known as "interconfluent hills." In the first stage depicted in the accompanying sketch the first cut-off is just beginning to develop. In the second stage it is completed and the first hill is detached from the spur. The present stage shows two hills with the possibility of a third one being detached later.

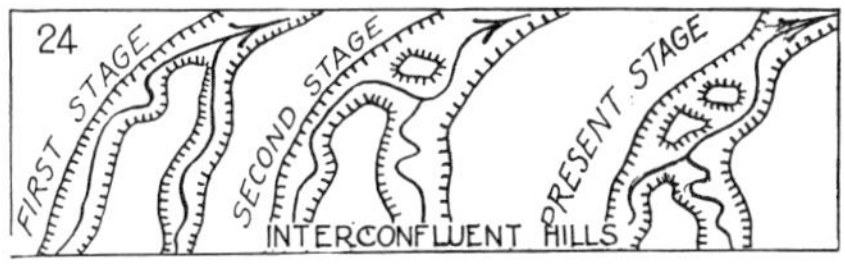

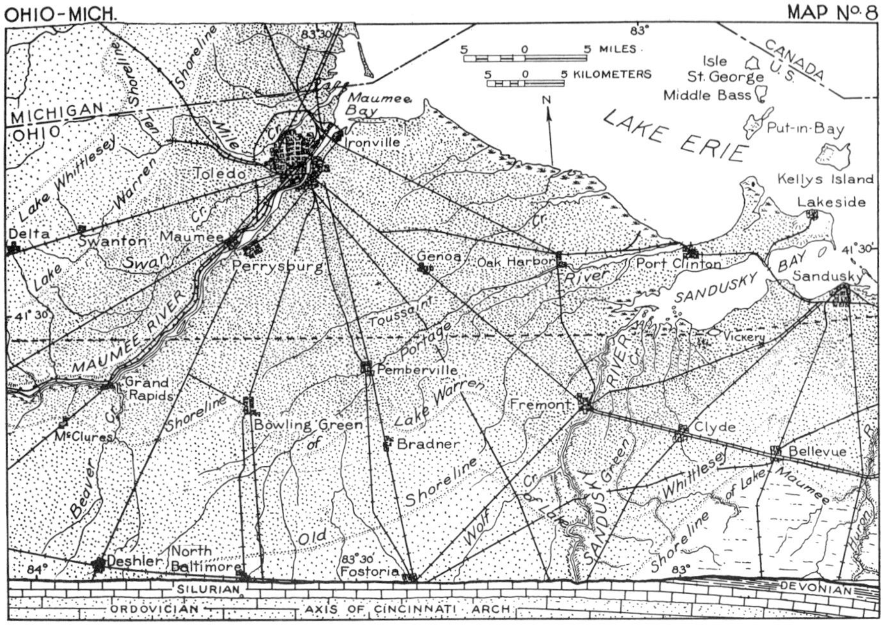

Route Map No. 8

THE SANDUSKY–TOLEDO REGION

The Embayed Shore of Lake Erie. Sandusky Bay, Maumee Bay and other similar features along the southwestern shore of Lake Erie are due to the submergence of valleys cut by streams. This submergence was caused by the tilting of the whole Great Lakes region in post-glacial time, the evidence for this being found in the fact that the old shorelines are no longer level but rise northward.

The airplane route skirts very close to the southernmost part of Sandusky Bay although at times it is likely that in going from Cleveland to Toledo the pilot will take his plane actually over the waters of the lake. The air traveller, however, might be interested to look for the Intermediate Landing Field at Vickery which is situated among cultivated fields and meadows very close to the shore of Sandusky Bay, at the point where the air route makes the turn around the lake. On one corner of the landing field is an arrow which points the way to Cleveland and New York.

The long gleaming white sand spit which extends northward and partly across the mouth of Sandusky Bay is a feature of interest easily seen from the air. Numerous other details of similar character may also be noted. At the head of Sandusky Bay the mouths of all the small streams which enter it have been drowned and the region is low and marshy (Fig. 28). Beyond Sandusky may be noted a number of islands in Lake Erie. The largest of these, Kellys Island, consists of Devonian limestone and is of geological interest because of the gigantic glacial grooves which the ice has gouged out of this yielding rock. With its islands and promontories the Sandusky Bay region has become one of the greatest inland resort sections of the United States.

Old Lake Beds. Between Sandusky Bay and Toledo the air route crosses the faint arch which constitutes the northern end of the Cincinnati anticline. The axis of this arch is the locus of the important Lima oil field which extends north almost

FIG. 28. Head of Sandusky Bay, Lake Erie, in its normal state, showing the submerged mouths of Sandusky and other rivers.

to Oak Harbor. As in the case of the Pennsylvania field, oil derricks are not to be seen here, and there is little to reveal the presence of oil to the observer in the air who can hardly detect the small pumps. Limekilns may occasionally be seen in this region, as this is a belt underlain by Silurian limestones. The bed rock, however, is almost everywhere concealed by a thin cover of old lake sediments. This part of Ohio is therefore extremely flat. Railroads and roads run for many miles without a curve (Fig. 29).

The Maumee River. At Toledo the route crosses the Maumee River, a stream which has a remarkable drainage pattern, most fully appreciated, perhaps, while

FIG. 29. The level till plains of northern Ohio, a view northward toward Lake Erie from near the western end of Sandusky Bay.

crossing Map No. 9. Several of its tributaries, as for example Tiffin River and St. Joseph River (Map No. 9) flow in a direction almost opposite to that of their master stream. The explanation is that these smaller streams were once tributaries to the Wabash, flowing to the southwest, and that the Wabash moraine (shown on Map No. 9) near Fort Wayne (south of Map 9), blocked their way and caused the upper part of the Wabash drainage system to become reversed and flow into Lake Erie. Hence this splendid example of a barbed drainage pattern.[25]

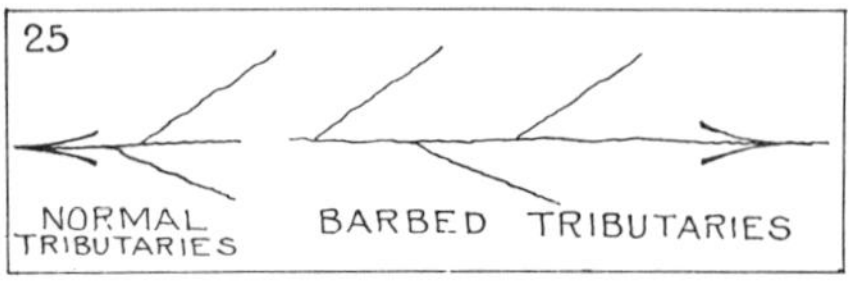

[25] Several of the features mentioned lie south of the area shown on Route Maps 8 and 9. A barbed tributary makes an extremely sharp angle with the main stream, resembling the barb on a spear.

This condition almost always means a reversal of drainage in the case of the main stream.

OHIO - IND.-MICH: MAP No. 9

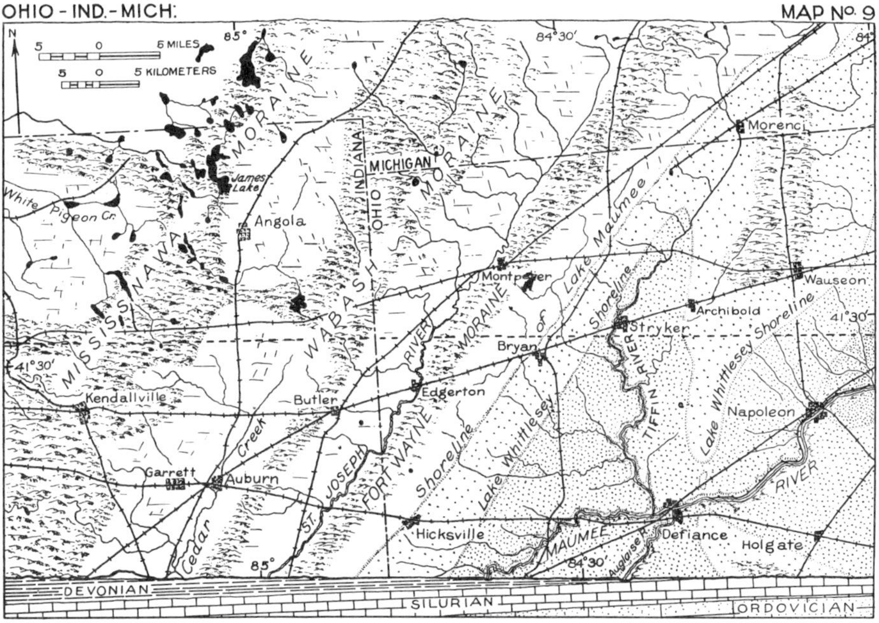

INDIANA - MICHIGAN MAP No. 10

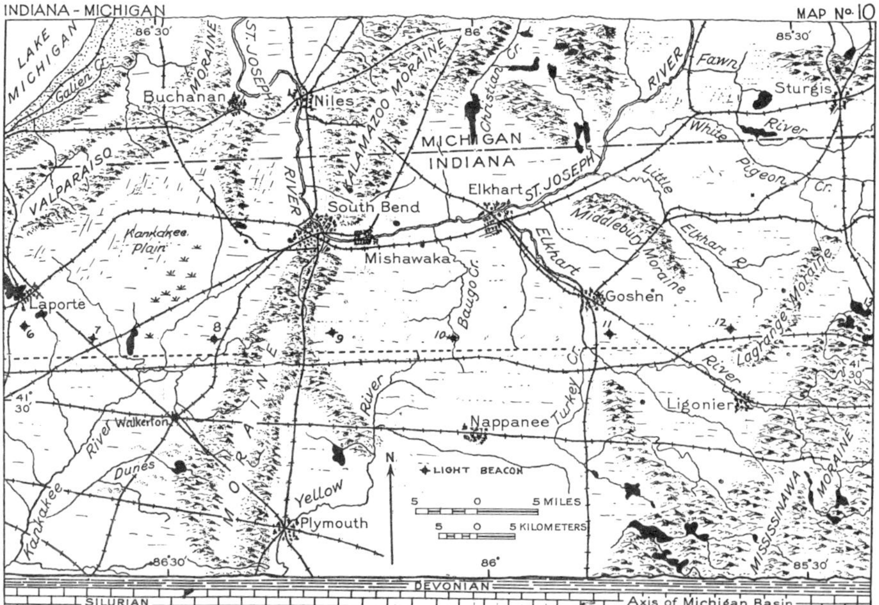

Route Maps No. 9 and 10

Route Maps No. 9 and 10

BRYAN, OHIO, AND SOUTH BEND, INDIANA, REGION

The Glacial Features of the Great Lakes Region. Northwestern Ohio and northern Indiana comprise a section of country which, to the air traveller, seems to have no diversity. Most of it looks like a vast checkerboard of rectangular farms (Figs. 30 and 31). Nevertheless, the air route traverses several different belts which

Fig. 30. The plains of northern Indiana, on the outskirts of Chicago. Hammond, Indiana, lies in the foreground. The smoke in the distance, beyond Calumet River, comes from the factories in East Chicago. In the distance is Lake Michigan.

can be distinguished by the discriminating eye. Fig. 32 (as well as Fig. 94) shows the arrangement of the moraines at the end of Lake Erie and of Lake Michigan.

Fig. 31. The plains of northern Indiana, a view looking northward toward South Bend which is situated on St. Joseph River, just visible in the distance. The main roads in this region are one-quarter mile apart.

Two lobes of the last ice sheet (known as the Wisconsin period of glaciation because of the numerous glacial features found in that state) came down through these lake basins as far south as central Indiana. As the ice retreated, or melted back, in stages, moraines were deposited around the front of the lobes, hence the scalloped plan of the moraines. The route from Toledo to Chicago goes from the center of one ice lobe to the center of the other.

From the air the morainal belts may be distinguished from the intermediate strips in several ways. First, the moraines appear more wooded. Although farms and cultivated tracts are located on the moraines, nevertheless the woodlots are distinctly more numerous and larger than elsewhere. This is a feature easily seen from the air. In the second place, the moraines usually include many lakes and ponds (Figs. 33 and 34), whereas in the belts between the moraines such features are almost totally lacking. Third, the road system on the moraines is much more irregular than in the flatter country between, where an almost perfect grid system is to be noted.

Four Agricultural Areas of Ohio and Indiana. The passage across this part of northern Ohio and Indiana involves four different agricultural areas which are rather definitely related to the geological conditions. The *first agricultural area,* and the easternmost one, is the Erie Plain (makes up most of Map No. 8) drained by the

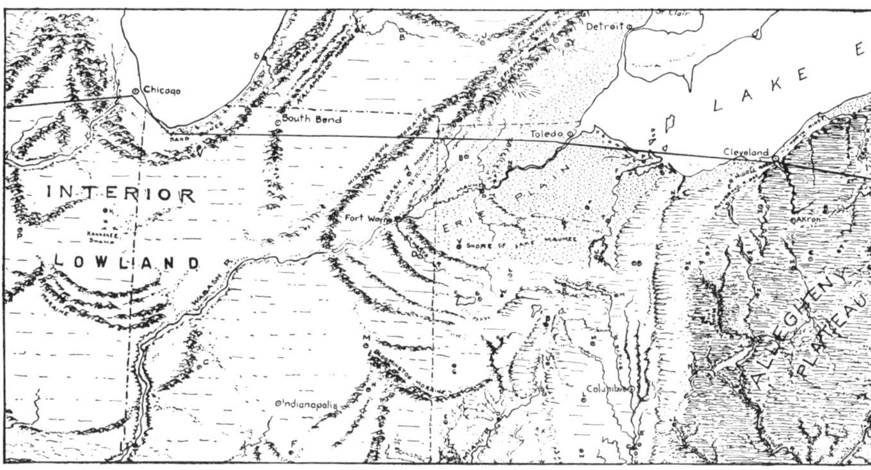

FIG. 32. Diagram showing the moraines and lake plains between Cleveland and Chicago.

Maumee River and its tributaries. This is a region of dark soil or muck, suited best to pasture lands, and to the growing of small grain. It is not so easily cultivated, hence corn is not so important here as farther west. In many places the bed of ancient Lake Maumee, with its rich black and poorly drained muck soils is devoted to the growing of onions and celery, and, near the larger cities like Cleveland and Toledo, to vegetables. The shoreline of Lake Maumee is crossed by the air route at Bryan, Ohio (Map 9). Its position can be identified by the fact that one of the diagonal roads of the region is located on it. Moreover, the soils of the country lying to the east are in general black and those west are buff-colored.

West of the Maumee lowland is a succession of moraines and plains extending west to the region of South Bend (Map No. 10). All of this area is comprised in a *second agricultural unit* where diversified farming is the rule. On the flat land cultivated crops are found, but the moraines, with their rougher topography, cause this region to produce more sheep than any other part of Indiana.

FIG. 33. LaPorte, Indiana, with large glacial lakes in the foreground. Lake Michigan in the distance.

The Moraines of Indiana. Just west of Bryan, St. Joseph River may be noted. This is one of the barbed tributaries of the Maumee, which, like the Tiffin, flows in a direction opposite to that of the Maumee itself. Then comes the Wabash moraine with several lakes included among its hills. Beyond, a few miles of open flat country intervene before the Mississinawa moraine (Fig. 34) is crossed just north of Kendallville (Map No. 9). This moraine is the largest and best defined moraine in Indiana, and is probably not surpassed in excellence of development elsewhere. Its surface is 150 to 300 feet above the country on either side, and its total thickness down to bed-rock from 200 to 485 feet. The surface is entirely occupied by deep, irregular,

FIG. 34. The Mississinawa Moraine in northern Indiana, north of Kendallville, showing numerous lakes.

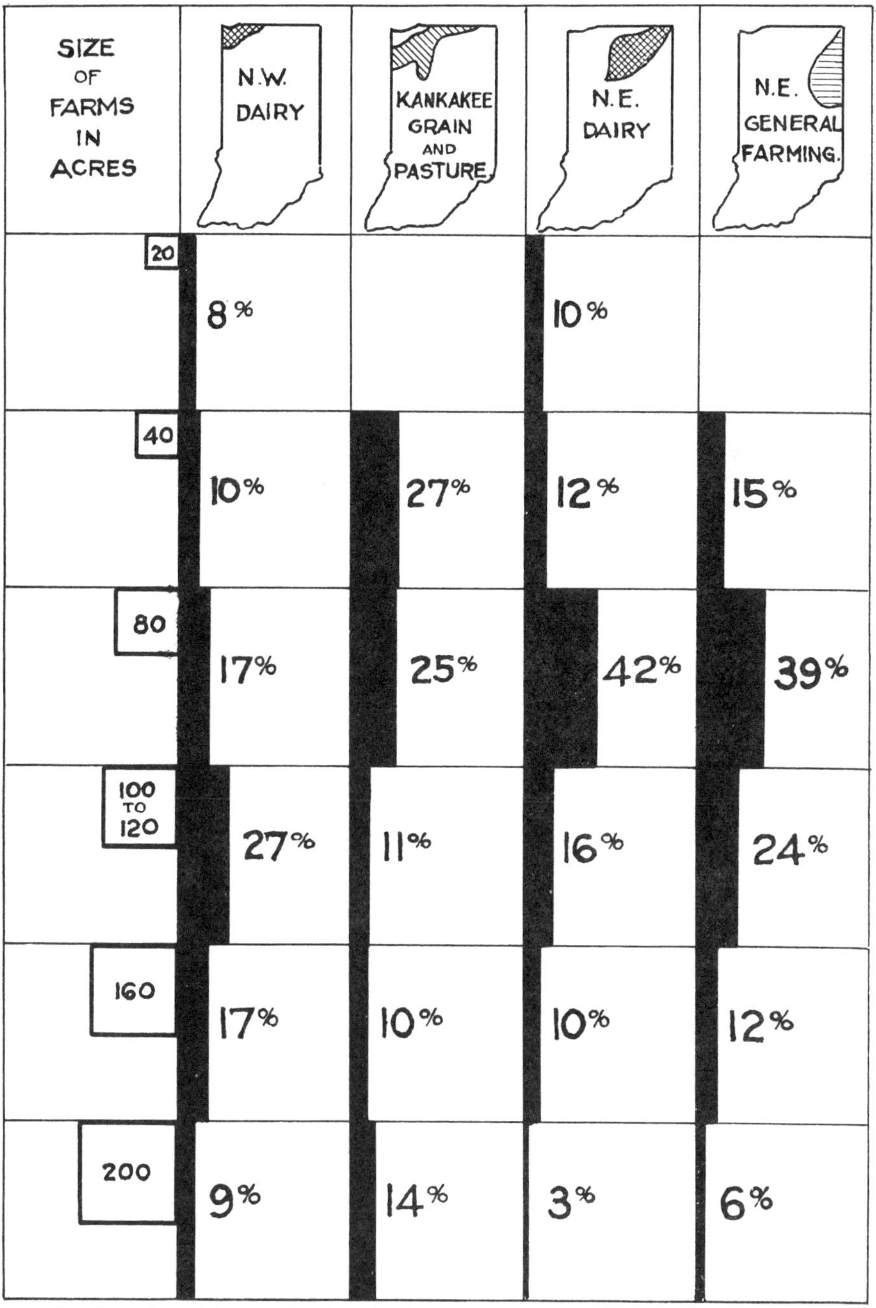

FIG. 35. Size of farms in the four farming zones of northern Indiana, showing percentage of farms of different acreage in each of the zones.

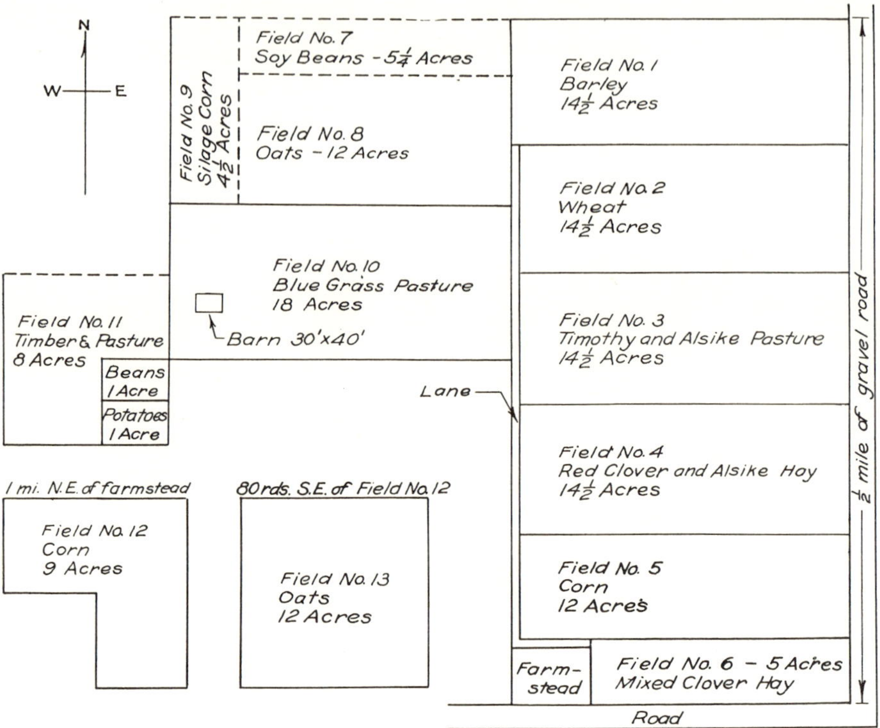

FIG. 36. Plan of an actual farm, near Walkerton, Indiana, showing utilization made of the different fields. This farm includes in all 146 acres. It is essentially a dairy farm but engages also in general farming.

elongated valleys, with narrow sharp winding ridges between, all in indescribable confusion. The roads through it are very crooked to avoid the marshes, and in every direction there are steep descents and ascents.

Scarcely more extreme and peculiar is the topography usually regarded as typical of terminal moraines, the "knob and basin." (See Note 22 on page 40.) It consists of confused groups of dome-shaped or conical hills, often as steep and sharp as the materials, usually sand and gravel, will lie, with hollows of corresponding shape between. The impression made is as if the material had been dumped from above and left as it fell, like gravel from a wagon. Throughout this morainic region the hollows or "kettle holes" are occupied by marshes or lakes and the extinct marshes or extinct lakes outnumber the living ones.

The intermorainic spaces within the area covered by the Saginaw ice-lobe, around Goshen, Elkhart, and South Bend (all on Map No. 10), is made up largely of outwash and gravel plains. North of the air route the gravel plains comprise probably half of the area. Southward, the intermorainic spaces are largely ground moraine[26] usually of an undulating nature, though some of it is quite flat.

The Kankakee Plain. West of South Bend and the Kalamazoo Moraine the route crosses the Kankakee lacustrine area which is a system of sandy lacustrine

[26] A lacustrine plain is a plain which was formerly the floor of an ancient lake. An outwash plain is a plain of sand and silt spread out in front of a melting glacier. A valley train is similar to an outwash plain but is more definitely confined to a valley. A till plain is made up of till or soil left by the melting ice and is therefore not assorted into sand or gravel like the outwash plains. Ground moraine is similar to a till plain but not so level.

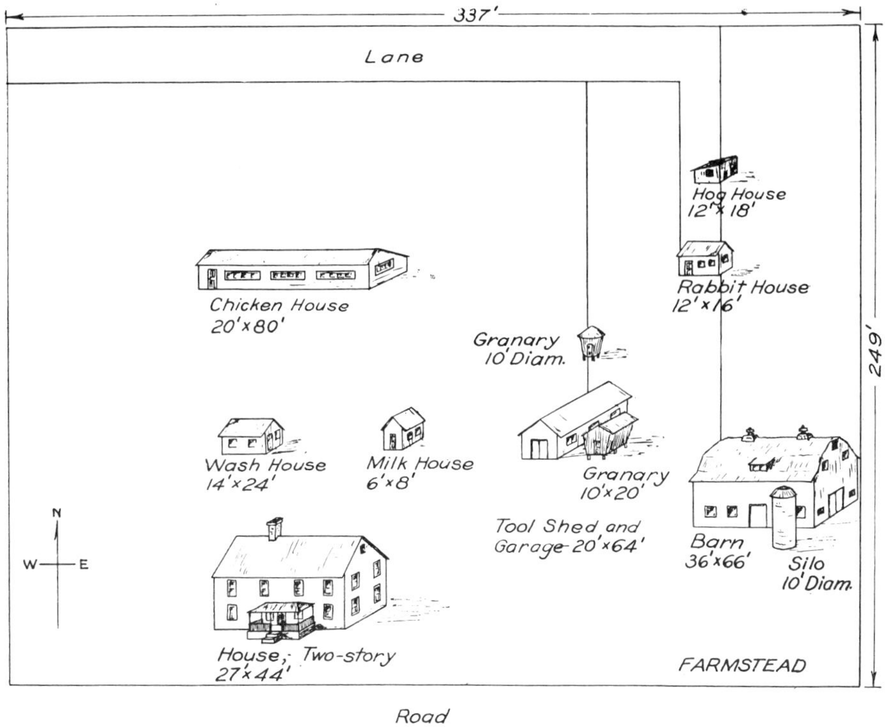

FIG. 37. The "Farmstead" of the farm shown in Fig. 36, enlarged to illustrate the arrangement of the farm buildings. These various buildings may be identified, from the air, on most of the farms in the middle west.

plains, outwash plains, valley trains, and local enclosed till plains associated with a great line of glacial drainage and ponding along the St. Joseph[27] and Kankakee Rivers. The belt is about 15 to 20 miles wide and embraces the Kankakee marsh and adjacent sand and gravel plains. This constitutes the *third agricultural unit* of northern Indiana, sometimes known as the Kankakee grain and pasture section. The greater part of this region is characterized by a thin deposit of somewhat ridged sand. Dunes are common. This sand was probably deposited in ponded or laked waters. The wide variety of soils in this area results in wide variation in type of farming. Much of this area has been surface-drained in recent years and much of it is still undrained. Corn and native pasture occupy a large portion of the farm area. This is the most important rye-producing section of Indiana. In this region there are large areas of muck land, much of which is given over to special types of crop farming. The two most important muck crops, mint and onions, are usually grown on these specialized farms. In some localities various types of truck farming have become important. Few sheep are found in this area and about one-third as many hogs as in the grain and livestock area which lies just south of this. The most important class of livestock is cattle. Beef cattle predominate because of the large areas of relatively poor native pasture unsuited to dairy herds.

The *fourth agricultural section* of Indiana is the northwest dairy belt adjacent to Lake Michigan. The region, however, is not well adapted to dairying because of

[27] This is not the same St. Joseph River, previously mentioned, which is tributary to the Maumee. This St. Joseph River flows northward across the corner of Michigan into Lake Michigan.

its sandy soil and poor pasturage. The proximity of Chicago, however, makes dairying profitable.

Notes on Indiana Agriculture. The observer flying over Indiana will find himself concerned more with the agriculture than with the geology of the region. Such questions as the actual size of farms, the arrangement and character of the farm buildings, the presence or absence of silos and windmills, the use of power machinery for plowing and harvesting, and the actual nature of the crops seen in the fields, will command his attention. To the layman it might seem that a few general statements could be made concerning these facts. Study of statistics shows, however, that there is more variation between the actual farms in a region than there is between farms in different regions. While the sum of all the statistics brings out the characteristics of a region, the individual farms do not each present an epitome of the region as a whole.

The accompanying table (Fig. 35), for example, shows that in regard to size of farms, there is great variation in each of the regions. Nevertheless it is noteworthy that the big Kankakee pasture area has a greater proportion of farms of large extent than do the more diversified farming regions lying to the east. Most farms are between 40 and 160 acres in size, with 80 acres predominating above all others in the eastern part of the state. This represents one-half of a quarter section or one-eighth of a square mile. Such a farm would probably be about half a mile long and one-quarter mile wide. About one-third of the farms have windmills. The others use gas engines and electric motors.

Figures 36 and 37 illustrate the layout of an actual farm in Indiana, lying almost under the air route, south of South Bend.

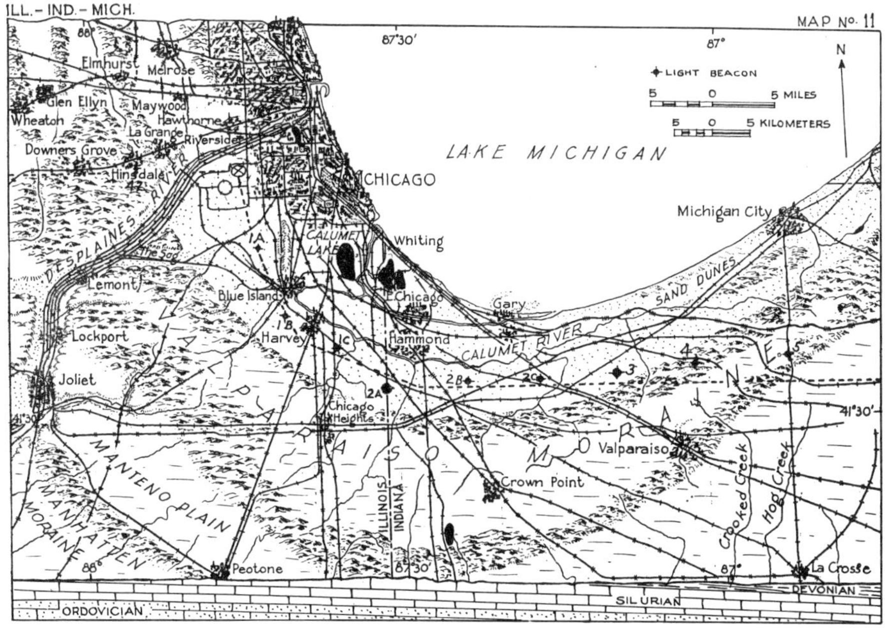

Route Map No. 11

THE CHICAGO REGION

The Chicago Plain. The city of Chicago (Fig. 38) is situated on a low, strikingly flat, crescent-shaped plain bordering the head of Lake Michigan. It attains a width of 12 to 15 miles in its widest part. From the shore of Lake Michigan, which is about 580 feet above sea-level, the plain rises gradually to a nearly uniform height at its inner margin, 640 feet above tide or 60 feet above Lake Michigan. This plain is the bed of former Lake Chicago, the glacial lake of slightly greater size which preceded the present Lake Michigan.

West and south of this plain is an elevated belt of more or less strongly undulating topography having a maximum altitude of 200 feet above Lake Michigan. This is the Valparaiso moraine, named after the city of Valparaiso, Indiana (on Map 11), through which it passes. It lies almost under the air route, and averages about 15 miles in width.

The Des Plaines River, between La Grange and Joliet, cuts across this moraine in a flat-floored valley a mile or so in width and 30 to 150 feet deep. At its lakeward end the bottom of the valley is continuous with the Chicago Plain and is less than 15 feet above the level of Lake Michigan. Tributary to this valley is a second valley of similar size which joins it on the east and in a like manner comes out on the Chicago Plain at a low level. This is called the "Sag." These two valleys served as the outlet for the waters of Lake Chicago when it discharged by way of the Illinois River to the Mississippi.

The Des Plaines River Valley. Inasmuch as Lake Michigan now belongs to the St. Lawrence drainage system and the Des Plaines to the Mississippi system, a continental divide may be said to pass through the Chicago Plain. This plain was formerly drained lakeward by the north and south branches of Chicago River, but extensive artificial modifications have been made which in a way reestablish the earlier condi-

Fig. 38. The City of Chicago viewed from above Lake Michigan. In the center of the picture is the Sanitary and Ship Canal. The main street upon which most of the tall buildings are situated is Michigan Boulevard.

(Photo by Chicago Aerial Survey, furnished by United Air Lines)

FIG. 39. The Municipal Airport at Chicago. Situated on the old bed of Lake Michigan. One of the beach ridges cuts across the far corner of the field.

tions of glacial time. The Sanitary and Ship Canal, and the Illinois and Michigan Canal now connect the Des Plaines River with the Chicago River. The latter stream has been dredged so that its flow is now reversed and the waters of Lake Michigan again flow southward into the Des Plaines. This great channelway, occupied by Des Plaines River, two canals and two railroad lines, is a conspicuous feature, best seen by the air traveller as he passes over it upon continuing his journey westward from Chicago.

The Beach Ridges of Ancient Lake Chicago. In the eastern part of the Chicago area, lying between the Valparaiso moraine and Lake Michigan is a series of beach ridges (Fig. 39) marking different stages in the elevation of Lake Chicago. These ridges (fairly well shown on the Toleston and Calumet sheets of the U. S. Geological Survey), rise so little above the level of the Chicago Plain as hardly to affect the courses of the numerous railroads entering that city from the east. But they do cause the Calumet River to double back on itself for a distance of 22 miles. Calumet Lake (Fig. 40), a conspicuous landmark from the airplane, as well as Wolf Lake,

FIG. 40. The outskirts of Chicago showing Calumet Lake and Calumet River on the old lake plain of Lake Chicago.

Lake George, and others which have been filled or drained, are parts of the bed of Lake Chicago, occupying swales between the smaller beach ridges. The most elaborate set of ridges and swales lies just south and west of Hammond, Indiana. Ninety of these ridges have been counted, ranging from three to ten feet in height, separated by narrow marshy belts. The larger beaches are covered with dunes. Dunes are also prominent along the present shore of Lake Michigan, notably on its eastern side because of the prevailing westerly winds. Owing to their height of 100 to 200 feet they are easily seen from the airplane flying eight or ten miles south of them.[28]

The City of Gary, Indiana. The air traveller approaching Chicago from the east will realize from afar that he is nearing a city of vast industries. The strip of glistening sand dunes along Lake Michigan comes abruptly to an end at Gary, In-

[28] This sand dune section of Indiana has been set aside as a state park. The dunes, however, continue northward for a hundred miles or more along the eastern shore of Lake Michigan. Dunes do not occur on the western side of the lake because the sand is blown into the lake by the prevailing westerly winds.

diana. Close to the lake front may be seen the forest of smokestacks which enable Gary to justify its claim to have the largest steel works, the largest tin mills, the largest rail mills, and the largest cement plant in the world. In the fall of the year 1905 the United States Steel Corporation decided to build a steel works in the Chicago district. Located midway between the ore beds of the Lake Superior region and the coal fields of Pennsylvania, in close proximity to markets and labor supply, with excellent water and rail transportation facilities, with unlimited water supply, and with large tracts of undeveloped land available, the site was ideal for the purpose. The present population of the city is over 100,000. The most striking landmarks visible to the air traveller are the mills of the Illinois Steel Company near the lake shore and the two wings of the 12-story Hotel Gary dominating the center of the city.

Just beyond Gary lies the city of Hammond, Indiana, with a population of 65,000, another very active industrial center. The Hammond Airport, situated on the Indiana-Illinois boundary, is one of the largest level airports in the United States. At this airport is located a radio marker beacon which marks the change in course between that leading to the Chicago Municipal Airport and the one leading eastward to Toledo and Cleveland.

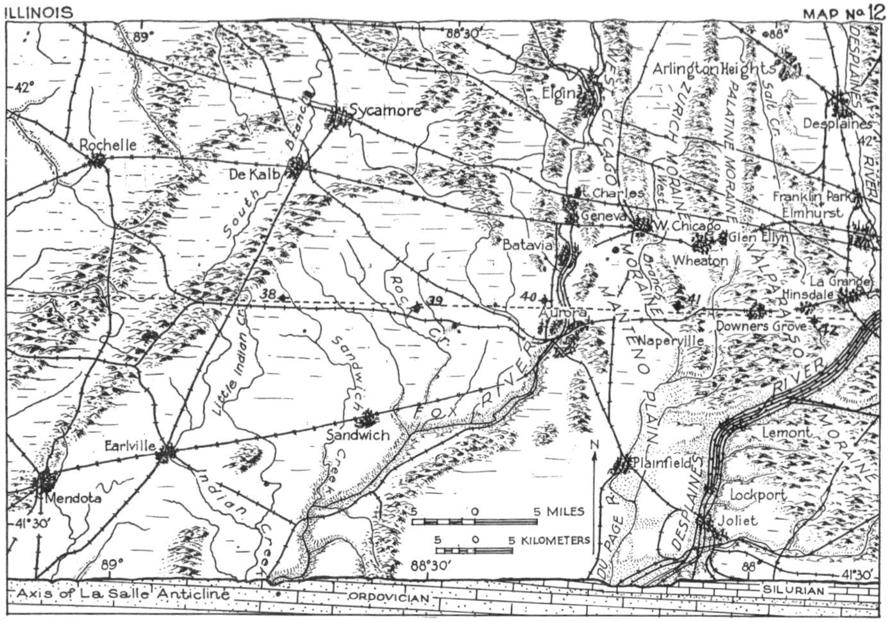

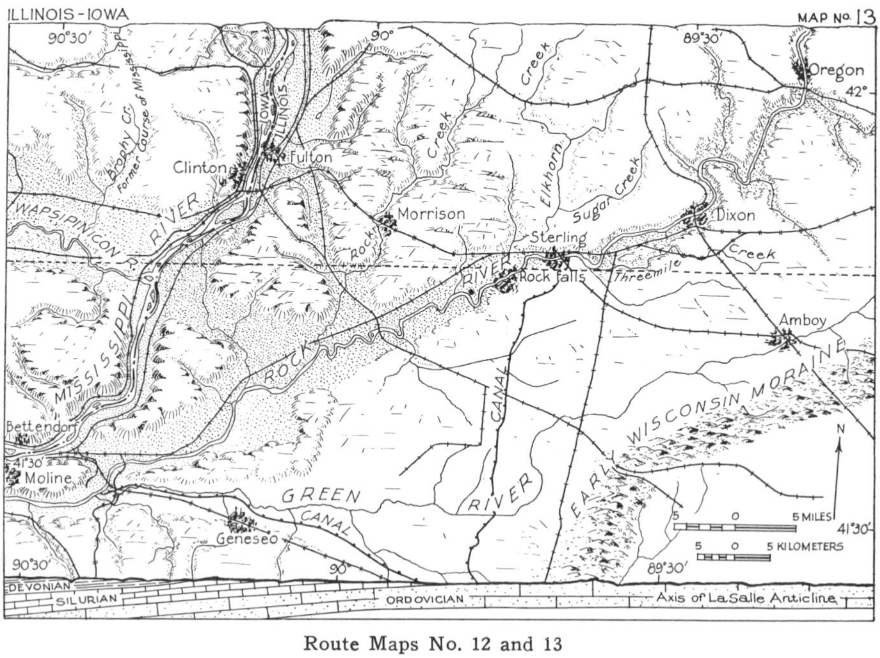

Route Maps No. 12 and 13

Route Maps No. 12 and 13

CHICAGO TO CLINTON

The Three Topographic Types of Northern Illinois. The region lying between Chicago and the Mississippi River includes three types of country. The eastern third of this region, as far as DeKalb (Route Map No. 12), is made up of a series of alternating moraines and outwash plains of the most recent glacial epoch. Lying west thereof, in the Rock River region (Map No. 13), are till plains and obscure moraines of earlier glacial time. Adjacent to the Mississippi River are maturely dissected, loess-covered uplands which grade northward into the Driftless Area.[29] The line of flight everywhere traverses glaciated country. The topography of the moraines is at best very obscure when seen from the air. The moraines reveal themselves by the number of trees and woodlots, the irregularity of the roads and fields, the small per cent of cultivated ground, and the occasional ponds and marshes,—all of this being in contrast with the flat till sheets and outwash plains lying between the moraines. The Rock River traverses a region of older drift which has been but slightly eroded. Near the Mississippi River, bed-rock is close to the surface and the country is strongly dissected.

The Three Farming Types. These three topographic types, the moraines, the till plains, and the uplands, correspond roughly with three general farming type areas, which to the air traveller offer certain recognizable aspects.

The morainal area near Chicago is a dairying region. Chicago receives a large part of its fluid milk supply from this district. Consequently this area produces a large percentage of valuable roughage and the light concentrated feed grains to be fed to the live stock, principally dairy cattle. The favorite breed of milk cattle here is the black and white Holstein which to the air traveller is easily recognized and distinguished from the brown and red beef cattle so common in the western part of Illinois. In this area one-quarter of the total farm land is in corn and almost all of this is cut for silage. Practically every farm has its silo.[30] One-half of the total farm land is used for oats, hay, and pasture crops, all of which indicates the tremendous demand live stock makes upon feed crops. In this area, comprising about one-twelfth of the state of Illinois, are to be found one-fourth of the dairy cattle. Poultry raising also ranks very high as a farm activity but only a few hogs are raised here. In the close proximity of Chicago large amounts of vegetable crops are grown. Between 50 and 75 per cent of the farmers' income in this entire region comes from dairy products alone.

In the second area, drained largely by the Rock River, mixed livestock farming is the prevailing type (Fig. 41). Thirty per cent of the land is in corn but only one-quarter of it is cut for silage, the rest being fed to hogs. Close to one-half of

29 A maturely dissected region is one with many streams. The upland surface is interrupted by many small valleys and gullies. The loess cover on the upland is a deposit of wind-blown silt which has accumulated since glacial time. It varies from only a few feet in thickness in Illinois to as much as 100 feet near the Missouri River. The "Driftless Area" includes part of Illinois and Wisconsin which was entirely surrounded by the ice sheet but was never actually covered and hence is without any glacial deposits.

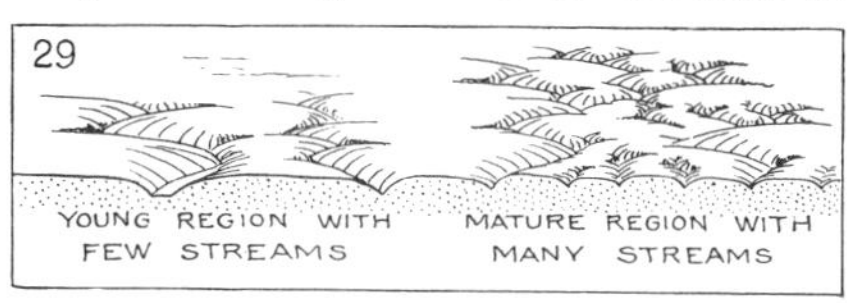

30 A silo is a cylindrical structure without windows, about 10 to 12 feet in diameter, and standing as high as the barn. Into this building chopped corn stalks are put to be used as cattle feed during the winter months.

(*Photo by Daily Pantagraph, Bloomington, Ill.*)

FIG. 41. Livestock farm at Sibley, Illinois.

the farmers' income is from hogs and less than one-quarter from dairy sales. Many farms in this belt do not have silos.

Adjacent to the Mississippi River practically one-third of the farm land is in pasture, more than in any other part of Illinois. This is due to the relatively great amount of rough, broken topography.

As in the second belt, hogs and cattle furnish most of the income, only 10 per cent coming from dairy products.

The Wapsipinicon Valley. Beyond the Mississippi River the route follows very closely the broad flat-floored valley of Wapsipinicon River. This valley, together with the broad valley of Brophy Creek joining it from the north, served at one time as the course of the Mississippi River when its present position was occupied by the ice sheet. This was during one of the earlier ice invasions, the Illinoian epoch. Westward from this point the Mississippi crossed the low divide into the valley of Cedar River (Map 14) and thence flowed southward, regaining its present course near Keokuk, Iowa, which lies south of the area depicted on Route Map No. 14.

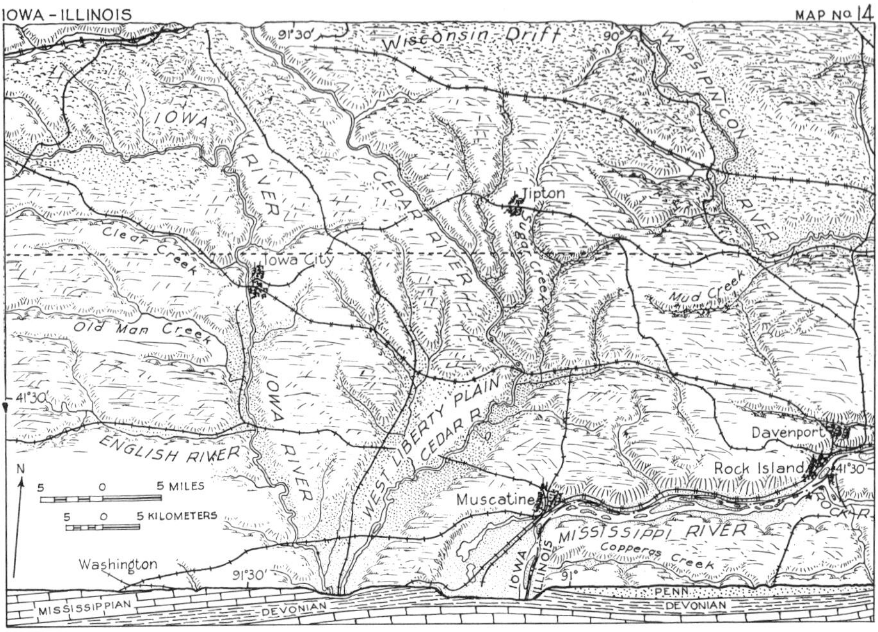

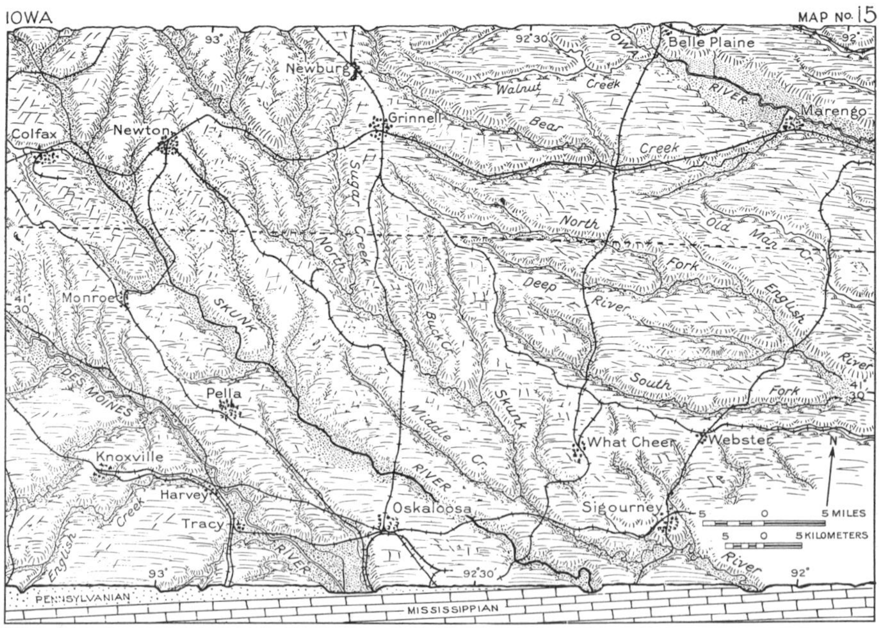

Route Maps No. 14 and 15

Route Maps No. 14 and 15

IOWA CITY TO DES MOINES

The Prairies. Between Iowa City and Des Moines the air route crosses part of Iowa which is covered by the older sheets of glacial drift. There is a certain amount of loess[31] covering this area also. Because this region has been exposed longer to erosion, the dissection of this part of Iowa is therefore greater than in north central Iowa where the most recent glacial deposits occur. Near the large rivers there are numerous ravines and gullies, usually supporting a growth of trees and thickets. Most of the country, however, is open, rolling farm land. The woodland areas are far less numerous and very much smaller than those of Indiana and Ohio. The broad divides between the streams were originally prairies, that is grassland without timber. In Iowa the prairie formerly covered the greater part of the surface of the state. Less than one-eighth was in forests and thickets bordering the streams. The prairies or grasslands were variable in topography, for they occurred in alluvial valleys, upon flat drift plains, on abrupt slopes, indeed upon all types of topography in the state from the flattest to the most broken. Neither were they restricted to particular geological formations, for they were found, and still exist to a limited extent, in their primitive condition, upon every kind of formation which reaches the surface, not only in Iowa but in the entire Mississippi valley.

Prairies are striking not only because of the absence of trees, but because they are marked by the presence of a flora able to withstand considerable drought. The primary cause for the treelessness of the prairies and their distribution in a given area is exposure to evaporation as determined by temperature, wind, and topography. On protected and shaded slopes forests develop, but on the exposed uplands it is difficult to raise trees, even artificially. Orchards are very rare indeed in this region.

The air traveller, looking down upon this landscape of almost continuous farms can, in his imagination, picture this region when the white man in his westward advance first beheld the prairies in all their splendor, and all their monotonous magnitude. These prairies presented varying aspects. The early settler avoided them at first in part for the reason that he thought them not fertile because treeless, and in part because they did not furnish the much needed building materials, fuel and water; and as his experience increased, there were added to these reasons the menace of the prairie fires and the terror of winter storms.

"Both of these dangers have practically disappeared with the settlement of the prairies and the planting of trees for shelter. The old-time blinding, bewildering blizzard of the prairies has lost its horror, and, though it may still cause personal discomfort, it no longer menaces the safety of its hapless victims.

"Even in winter the prairie was often attractive, for the storms subsided, and by day the sun-lit sea of snow sparkled with countless ice-crystals, and by night it rested in impressive silence under the star-spangled sky.

"With the opening of spring the ponds and lakes were gilded with the water crowfoot, and the hills and higher prairies were dotted with the early pasque-flower, the prairie violet and a variety of spring flowers. The broad prairies were swept by great whirl-wind clouds of golden plovers; the long-billed curlew hovered between earth and space; the bobolink and the marsh blackbird made the air ring with their songs; the mournful boom of the prairie chicken resounded everywhere; and soon countless nests were occupied by wild geese, ducks, and prairie hens on all sides, giving promise of new life in untold numbers to enliven these prairies in a fashion which will never again be known to this or coming generations of men. Soon the grasses covered the surface with a great carpet of green, painted with puccoons, prairie phlox and other flowers of late spring.

[31] Loess is a fine wind-blown deposit of silt or dust, often many feet in thickness; very common in the Mississippi and Missouri valleys.

"But the real rich beauty of the prairie was developed only after mid-summer when myriads of flowers of most varied hues were everywhere massed into one great painting, limited only by the frame of the horizon, uniform in splendid beauty, but endlessly varied in delicate detail.

"In the fall this in turn was followed by the rusty-red or brown expanse of drying grasses which portended the coming of the terror and the splendor of that scourge of the early prairie settlers, the prairie fires.

"Such were the prairies as the early settlers saw them. But the native prairie is fast disappearing before the army of homebuilders, whose invasion is everywhere followed by the disappearance of the native prairie flora, and the appearance, in its stead, of artificial groves, cultivated crops and introduced weeds."[32]

The Former Lake Calvin. South of Iowa City and hardly within easy view of the air traveller the valley of Iowa River is broad and flat-floored. The valley of Cedar River, one of its tributaries, has a similar form. These flat plains constitute the bed of former Lake Calvin, one of the earlier glacial lakes, which resulted when the outlet of Iowa River was blocked by the Illinois glacial lobe, which then covered that state.

The Lake Calvin basin is an extensive lowland surrounded on all sides by drift uplands rising above it to a height of eighty to one hundred feet. Its topography is that of a monotonous plain, and the courses of Iowa and Cedar Rivers are marked by various sloughs, abandoned channels, marshes and crescentic ponds.

Several sets of terraces occur in the Lake Calvin basin, the airport of Iowa City being situated upon one of them.

The Kansan Drift Plain. This region comprises the greater portion of the highland bordering the Lake Calvin basin, and consists of a maturely dissected upland. It is called "Kansan" because it is covered with glacial drift deposited by an earlier ice sheet which advanced into Kansas. Originally, at the time of deposition, it was a somewhat gently-sloping flat plain, but its surface today is deeply incised by an intricate dendritic system of valleys. As one approaches the larger stream valleys after flying over the interstream areas, a gradual and orderly change from the uneroded, flat, original drift plain to a rolling and finally rugged and rough type of country makes itself manifest. The region east of Iowa City between Iowa and Cedar Rivers reveals some of the flat erosional remnants not yet showing the effects of running water. Away from these uneroded remnants of the original plain and nearer the master streams the surface of the upland is more indented by the numerous ramifications of the arborescent drainage system. The broad-bottomed open swales with long gentle slopes are constricted to narrower, steep-sided valleys, the relief becomes greater until finally at the borders of the upland the streams have thoroughly dissected the region.

32 Shimek: The Prairies.

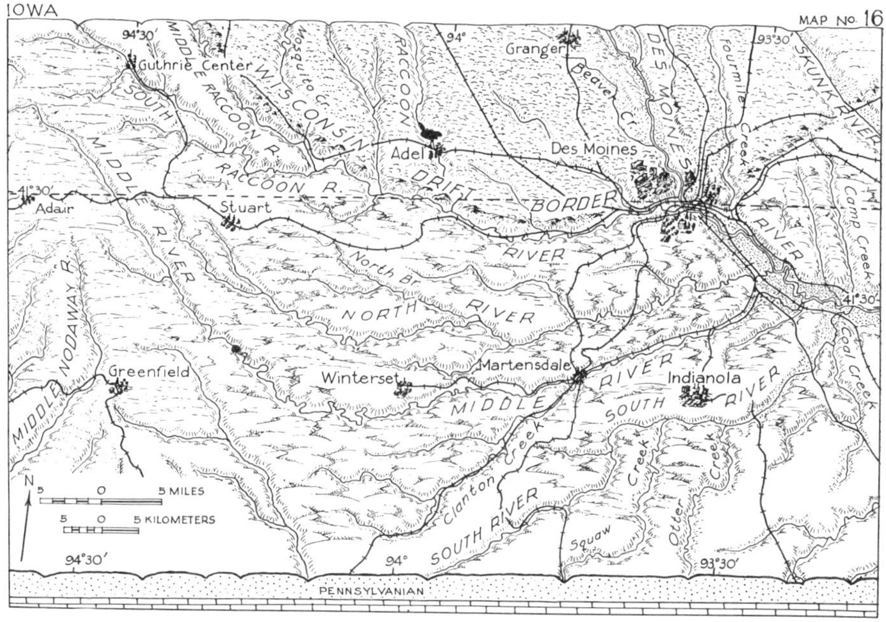

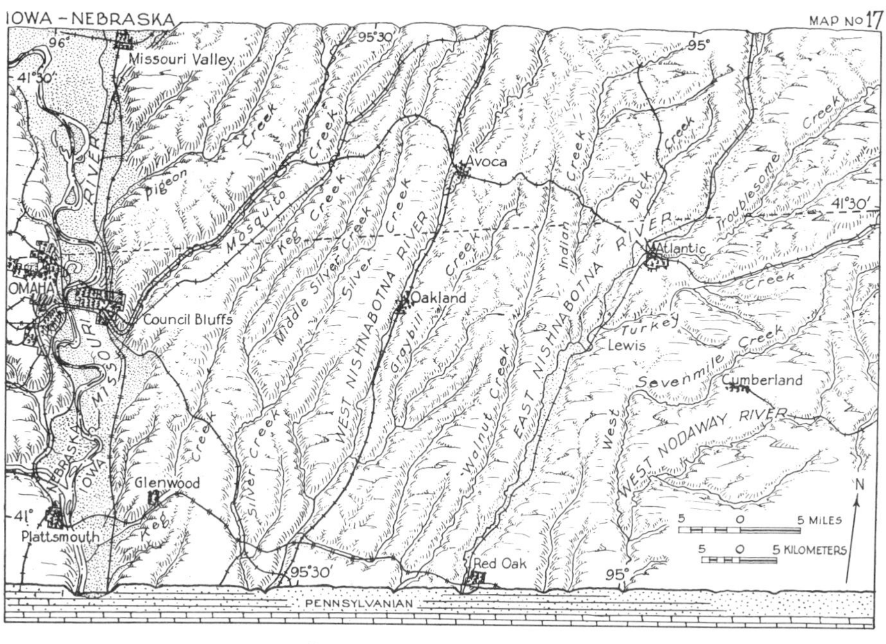

Route Maps No. 16 and 17

Route Maps No. 16 and 17

DES MOINES TO OMAHA

The Glacial Topography of Iowa. From Des Moines to Omaha, just as in the region east of Des Moines, the air route passes over part of Iowa covered by the Kansan, that is one of the oldest, glacial drift sheets.[33] North of the Raccoon River may be seen the southernmost part of the area invaded by a tongue of the Wisconsin ice sheet.[34] The Kansan drift area is in most places maturely dissected and has a strongly rolling to hilly topography. The area covered by Wisconsin drift is still very youthful and presents a flatter appearance than the older regions adjacent to it.[35] Near the Missouri River there is a heavy loess[36] cover which thins gradually eastward as far as Des Moines.

The Wisconsin Drift Area. The Wisconsin till plain still has a typical glacial topography. It has been but little modified by erosion, comparatively few streams cross its surface, and the terminal moraine, just north of Raccoon River, is nearly as distinct as when first formed. Wide expanses of prairie stretch to the horizon. Upon this generally level surface are numerous depressions, some of which are filled with water and form lakes and ponds. Others have been partly filled with silt and now contain beds of peat. Some of these lakelets or peat swamps may be found almost at the margin of the Wisconsin drift plain, close to the Des Moines and Raccoon Rivers, testifying to the slight extent which these streams have affected the topography left by the last ice-sheet.

The Kansan Drift Area. In contrast with the Wisconsin drift area, streams have been long at work on the Kansan drift sheet and have cut deep trenches into the till. While the skyline is almost everywhere level, the surface is scarred and gashed by rivers, creeks and ravines. The city of Des Moines stands just at the margin of the two drift areas. North of Des Moines the Des Moines River has exerted very slight influence upon the younger topography whereas south of the city its effect on the vastly more mature landscape is very noticeable. It has not only widened its valley but side wash has assisted it in smoothing down the valley slopes, and its many tributaries have penetrated headward into nearly all the surrounding prairies, which have been transformed into more or less flat-topped ridges and divides separating wide valleys. When the white man first settled in this region these valleys were covered with timber, which served as a protection against the wash of the slopes and bottom lands. The settlers' needs and their successors' greed and thoughtlessness have transformed many of these beautiful and useful timber tracts into barren wastes gashed by rains and trenched by ravines. With the clearing and cultivation of the hills and slopes throughout the area of Kansan drift there has come inevitably an increased wastage of the soil. Many of the hillslopes in southern Iowa now display, instead of that rich covering of black loam which had accumulated during ages of plant growth, yellow patches which show where the top soil has been washed away, leaving exposed the unmodified and therefore less fertile loess or drift. The flat floors of the larger valleys are made up of the finer materials which represent the surface wash from the fields and hillsides drained by their tributaries. These bottom lands, with their fine black alluvial silts are exceptionally fertile, thus recompensing in some degree for the loss sustained in other areas.

The Des Moines River. At Des Moines it may be noted that Des Moines River passes through the middle of the city, west of Capitol Hill. East of the city, however,

[33] The term "drift sheet" simply means glacial deposit.

[34] The Wisconsin ice sheet was the last one to advance over the country. The Kansan ice sheet was one of the first.

[35] The older regions have been more dissected by streams than the younger ones have.

[36] The loess, or wind blown silt, is thickest in the west whence the material came.

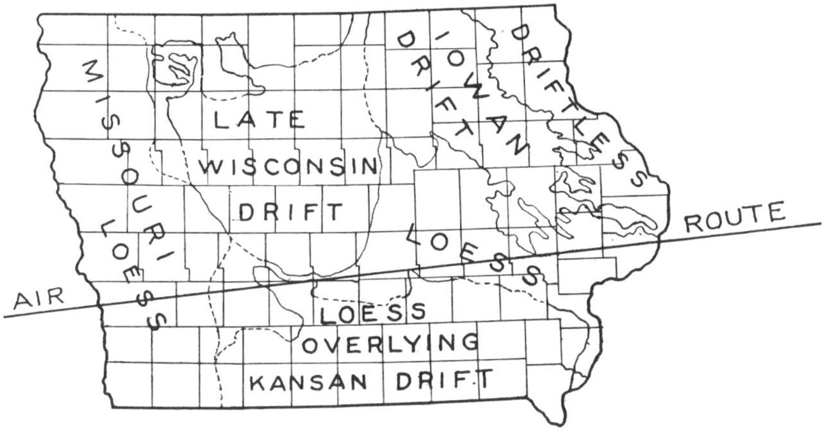

FIG. 42. Map showing the glacial deposits of Iowa.

there is a well defined valley which represents the pre-Wisconsin course of the river, that is the path used by the river before the last ice sheet encroached upon this region.

The Raccoon River. Raccoon River occupies a valley just outside, or to the south, of the Altamont moraine.[37] Its valley, which is followed several miles by the air route, is wide and flat, at least a mile across, and it contains immense quantities of sand. The north wall of the valley is high and rugged. Short, steep ravines cut into this wall and the neighboring upland, and give them a rough and rolling aspect. This north wall was covered with ice during the Wisconsin epoch and a mantle of drift was left upon it. The south wall is for the most part rather gentle, the ravines are not so steep, and the bluffs are less bold. It still retains its pre-Wisconsin features unaffected by the glacier's work.

The Agriculture of Iowa. The agricultural activities in Iowa reflect strongly the geological and topographic conditions. Figure 42 shows the distribution of the

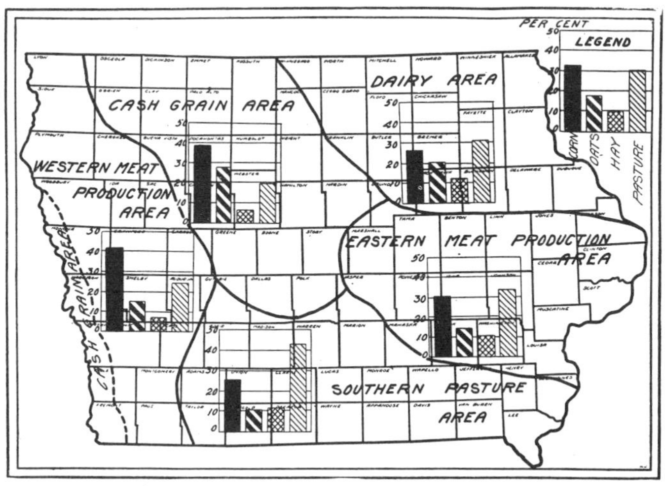

FIG. 43. The Type-of-farming areas of Iowa with their cropping systems. Note the significant differences in the proportions between the four important uses of Iowa farm land, corn, oats, hay and pasture.

[37] This moraine forms the Wisconsin drift border, shown in the northern part of Route Map No. 16. South of it is the older region of Kansan drift.

FIG. 44. The rolling, maturedly dissected plains of Iowa. All of the valleys, the ravines and steeper slopes are covered with trees but most of the interstream area is cultivated.

several glacial drift sheets. Figure 43 shows the utilization which is made of the farm land.

The cash grain area practically coincides with the Late Wisconsin drift plain. Corn, oats, and other small grains are the important crops. The flat land is conducive to extensive cultivation. It is the only part of Iowa where the percentage of pasture land is less than the percentage of oats. This small amount of uncultivated land is due to the good soil and the undissected fairly level topography characteristic of young drift plains. The landscapes in the region are laid out in a very regular checker-board plan, with roads following the section lines, this being another indication of relatively level and unbroken land.

The southern part of Iowa consists of the older Kansan drift, considerably dissected (Fig. 44), so much so that it is inadvisable to cultivate all the arable land

FIG. 45. The Missouri River, near Omaha, showing part of the flood plain, and in the distance the bluffs on the eastern side.

because of the gullying which results. Much is left, therefore, in permanent pasture. The Southern Pasture area which coincides with this region has three times as much pasture acreage as that devoted to oats. In this part of Iowa, the roads follow the divides.

Western Iowa, in spite of its fairly high relief, is the most important corn producing part of the state. This region is covered with a thick deposit of loess and is extremely fertile. Most of the corn grown here is fed to hogs; the fairly extensive pasture on the slopes supports cattle, with the result that this is the most important meat producing section of the state.

The Flood Plain of the Missouri River. The bottom lands, that is the flood plain, of the Missouri River are easily cultivated and have high fertility. They are, therefore, devoted to the growing of corn and smaller cash grains. However, the observer will note that on the flood plain of the Missouri River there is little if any livestock and consequently there are no fences. In fact, few people actually live there because of the poor, unsanitary living conditions. The land is subject to overflow and the drainage is poor. All of this discourages the raising of livestock because where livestock are raised people must be constantly on hand to care for them.

The steep, gullied bluffs, composed almost entirely of loess, rise above the flood plain of the Missouri River on either side. These are scantily clad with vegetation, and the almost vertical slopes reveal the yellow loess deposits, which in some places are almost 100 feet in thickness. On these bluffs and in the gullies may be seen shacks inhabited by so-called squatters who have small farms in this inhospitable environment. In other words, this is a marginal type of agriculture.

Throughout Iowa the traveller will notice that the valley slopes are maintained in permanent pasture, or else are forested if they are very steep; the flat divides between the streams are used for corn or other cultivated crops; the flat valley floors, if wide, are used for corn or other cash crops, and if narrow, for pasture land.

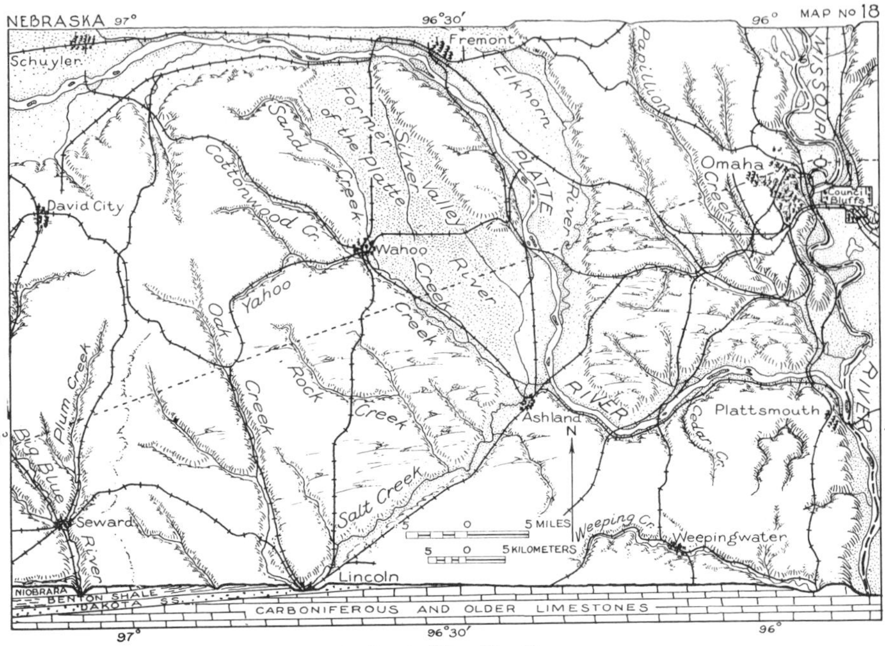

Route Map No. 18

THE OMAHA–LINCOLN REGION

The Bluffs of the Missouri River. The landing field at Omaha is situated on the flood plain of the Missouri River between Florence Lake and Cutoff Lake, two oxbow lakes,[38] formerly parts of the Missouri River (See Omaha sheet of the U. S. Geological Survey). On either side, the broad flood-plain of the Missouri, which here averages four miles in width, is bounded by prominent bluffs rising 200 to 300 feet above the river level (Fig. 45). These are especially prominent on the eastern side of the river and are seen in great detail by the air traveller because of the low elevation at which they are crossed, both when landing and taking off. The uppermost hundred feet of the bluffs is composed of loess, a fine yellowish silt deposited there by the wind. Because of its compactness it forms steep valley walls. Underlying the loess is a thick deposit of old glacial till and gravel. On the bluffs the occasional shacks of the poorer farmers constitute the only evidence of habitation.

The extremely muddy appearance of the Missouri River in contrast with the Mississippi strikes the attention of the traveller immediately. The Missouri carries more silt than any other large river in the United States. Each year 381 tons of dissolved and suspended matter are removed on the average from every square mile of country which it drains.

The Loess Plains of Eastern Nebraska. The entire region lying west of Omaha and north of Lincoln is a maturely dissected plain underlain by limestone of Carboniferous age. Overlying these limestones is a deposit of early glacial till. This

38 An ox-bow lake is a crescent-shaped body of water representing part of the former river channel, now deserted by the stream.

On almost all flat-floored flood plains, both large and small, ox-bow lakes may be seen.

in turn is covered with the wind-blown loess which constitutes the surface of the country for thousands of square miles in eastern Nebraska. The loess occupies more than one-third of the area of Nebraska, usually forming level plains which it underlies to a thickness of about 100 feet. The widest area is in the table-lands south of Platte River. The loess deposit is of a remarkably uniform texture, compact, but perfectly soft. Much of the water falling upon its surface is absorbed, but on steep slopes there is rapid excavation wherever a rivulet gets a fair start.

The loess is the product of a relatively long period of deposition at the close of glacial time. Much of it is wind-blown, but in central Nebraska it was deposited by the Platte River at a stage when the river spread as a wide network of streams over the divide south of its present course. One of the former valleys of the Platte may be seen about 35 miles west of Omaha, just north of Wahoo.

The eastern portion of the loess-covered area, well represented by the region between Omaha and Lincoln, is much dissected by present-day streams. This part of Nebraska is distinctly rolling, and even hilly. It is not unlike much of Iowa, and is a perfect example of a mature plain. All of the land is in slopes.

This is the important corn producing section of the state, although a great deal of wheat is also grown as is evidenced by the presence of grain elevators in every town along the railroads. There are no silos. Nevertheless cattle are important, both for beef and for dairy purposes. Hogs and chickens, too, may be seen in considerable numbers. This is part of the southeastern general farming area of Nebraska. It owes its prosperity to the fairly abundant rainfall and the high fertility of the loess soil.

Windmills are used on practically all the farms in this region, as they are throughout Nebraska. While most of the farm wells are shallow, some deep wells have been put down, notably at Lincoln, to a depth of over a thousand feet into the quartzites underlying the Carboniferous limestones.

In the eastern part of the area the towns are situated largely along the stream courses, which in general have fairly wide valleys. Lincoln, for example, on the Salt River, is situated where several tributaries of that stream converge.

Farther west, however, where the land is much less dissected, and where the loess constitutes a kind of plateau, the towns occupy the flat uplands.

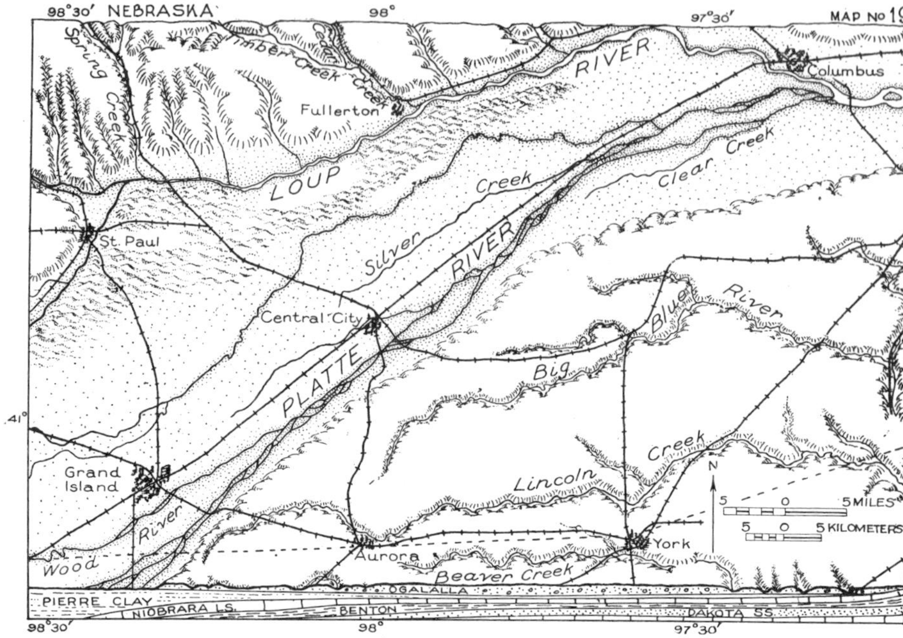

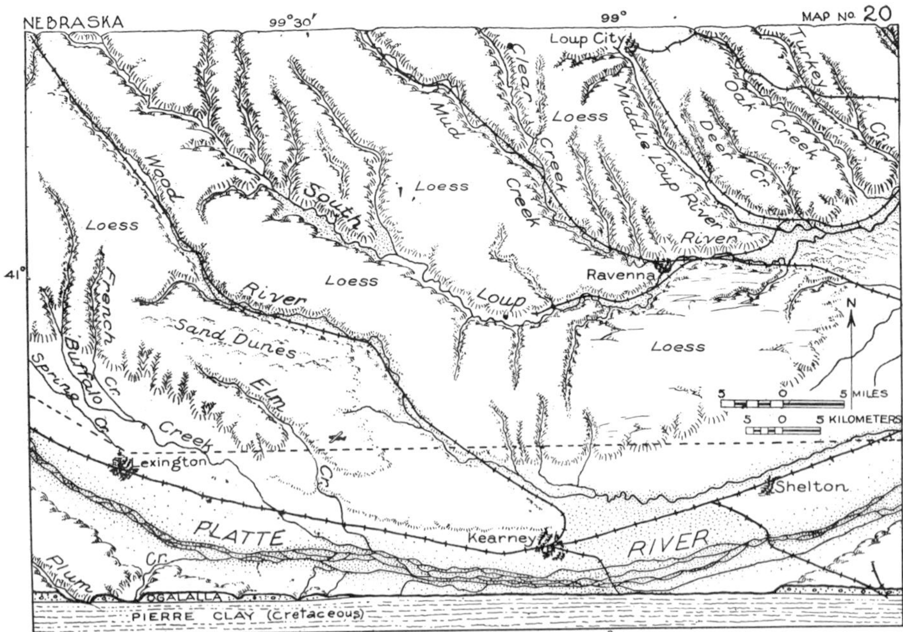

Route Maps No. 19 and 20

Route Maps No. 19 and 20

THE YORK, GRAND ISLAND, AND KEARNEY REGION

The Loess Plains of Central Nebraska. From Seward (see previous map, No. 18) westward past York, and on to the Platte River, the divides between the streams are extremely flat and undissected. These are loess-covered plains and constitute the most nearly level part of the state, more so than even the High Plains around Cheyenne. The eastern margin of these plains is marked by an obscure escarpment which runs approximately through Seward (Map No. 18) and represents the headward limit of erosion accomplished by the tributaries of Salt Creek.

The region between Lincoln and Omaha, however, is an older plain characterized by rolling topography. The belt lying west of the escarpment just mentioned is really the eastern part of the High Plains,[39] being underlain by the Tertiary gravel beds (called the Ogalalla formation) covered with loess.

In this flat part of Nebraska beef cattle are common and wheat is an important crop, as it occupies about one-fourth of the farm area. Corn, however, is the dominant crop, occupying nearly one-third of the farm lands. This is called the southern cash grain and live stock area. The rainfall averages about 28 inches per year, as compared with 32 inches or more in the easternmost part of the state. Moreover, the rainfall is very variable, and this is especially true during the summer months, with the result that the yields of crops like corn vary from year to year as much as 41 per cent from their long-time average. The variability of wheat is very much less because of its winter habit of growth.

The location of the air route across this part of the country depends upon whether a stop is made at Lincoln. In that case the route is practically over York and Grand Island. Otherwise it is 20 to 30 miles farther north.

The Valley of the Platte River. The Platte River valley (Fig. 46) is the chief feature of interest in this portion of the trip. At Columbus (Map No. 19) it is joined by the Loup River (Fig. 47) which, with its tributaries, drains the sand hill region of central Nebraska. The Platte River is one of the few streams which cross the entire width of the Great Plains.[40] It flows on a broad level flood plain 10 to 12 miles wide, above which rise bluffs 100 feet or more to the level of the High Plains. The river itself consists of many small irregular streams among the sandbars, forming a lacework of channels, whose pattern changes with every flood. During the dry season there is little or no water in the Platte as the water is diverted farther upstream for irrigation.

The Platte River valley is the locus of the larger towns of central Nebraska. In contrast with the drier plains on either side this valley is a fertile oasis. The raising of sugar beets, alfalfa and other crops which respond well to irrigation is important. The fields of alfalfa are readily distinguished from aloft because of their deeper green, as contrasted with the much paler green of oats and other grains.

Vast herds of buffalo used to roam this region, each spring migrating from the plains of Kansas and across the Platte to the richer grasslands farther north.

The Loup River. Beyond Grand Island (Map No. 19), and north of the Platte River, the air route traverses a broad loess-covered plain which in general resembles that lying east of the Platte, but it is much more dissected, owing to the presence of the Loup River with its numerous tributaries.

Corn is the most important cultivated crop in this section, but as one rides westward he will be impressed with the greater and greater extent of the area devoted to pasture. The small farms common farther east now give way to big ranches, many of them a square mile in area. There is no regular road system, due

[39] The High Plains comprise a vast territory in western Nebraska, Colorado and adjacent states, of very level land made up largely of gravel beds spread out as alluvial fans from the foot of the Rockies in Tertiary time.

[40] The others are the Missouri, farther north, and the Arkansas and the Canadian farther south.

FIG. 46. The Platte River in central Nebraska, east of North Platte, where it flows in several separate channels, over a wide flood plain.

to the unevenness of the topography. The slopes of the hills are gullied or windblown. Some sand dunes appear. Cattle are numerous. Windmills, isolated from all buildings, may be seen in the hollows between the hills. Here the water is stored in small ponds for the cattle. Occasionally a main road of the region is laid out perfectly straight for miles but it goes up and down over the hills.

In the open lower and flatter land adjacent to the streams, as for example in the valley of Wood River, sand dunes are absent and the region is well cultivated. The road system, too, in such areas, is regular. Much of the land is under the plough. Four-horse teams may be seen at work in the fields. Some of the uplands, on the other hand, as for example the region lying between Wood River and the Platte River north of Lexington (Map No. 20), are covered with sand dunes and the cultivation is patchy.

FIG. 47. The Loup River in eastern Nebraska, near its junction with the Platte.

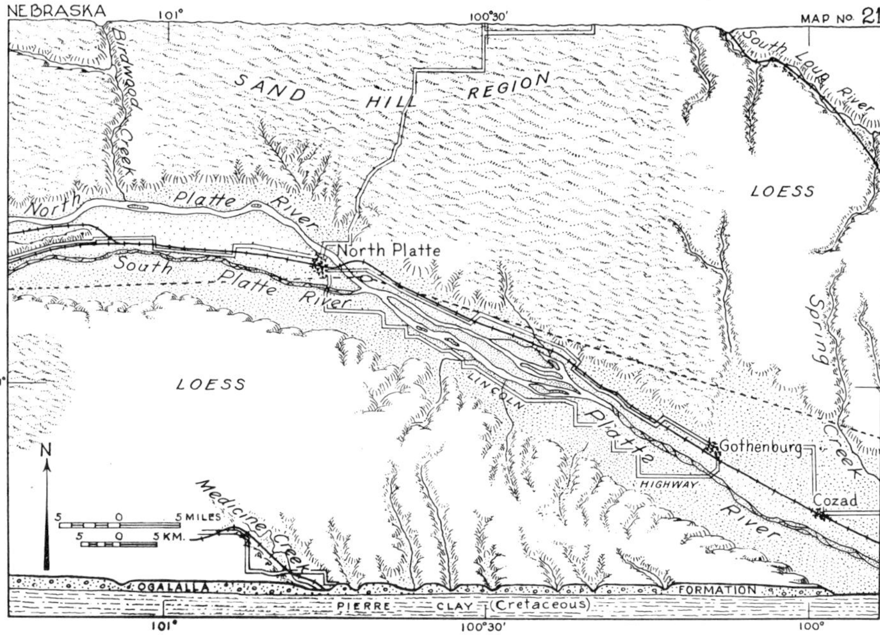

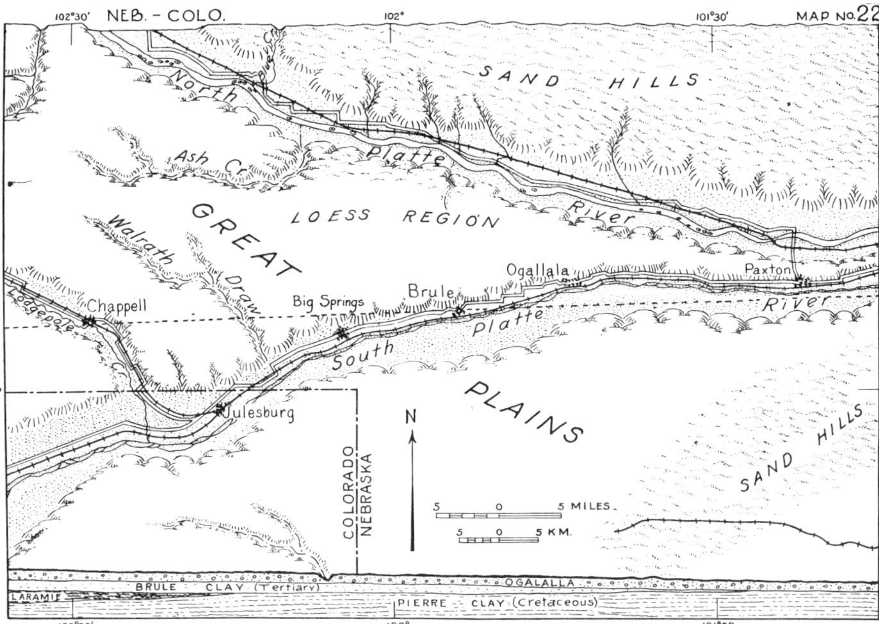

Route Maps No. 21 and 22

Route Maps No. 21 and 22

THE NORTH PLATTE AND OGALALLA REGION

The Nebraska Sand Hills. For 30 or 40 miles east of the city of North Platte (Map No. 21) the air route traverses the margin of the vast Sand Hill district which comprises over 24,000 square miles in northwestern Nebraska. It is really a transitional zone between the main sand-hill section and the loess-covered plains to the east. As it is the only part of the route which actually crosses some of these hills it may be well to note their topographic aspects.

The broad undissected cultivated tracts making up the loess plains north of Gothenburg (Map No. 21) give way westward to uncultivated billowy areas of sand hills, some grass-covered, others with great gashes and blow holes glistening in the sun. There are no trees whatever. Flat meadows of rich green grass appear between the dunes. Occasional windmills supply water to the very small round ponds toward which cattle trails lead from all directions. As far as the eye can see, sand-hill country of this kind stretches away to the north, while southward the valley of the Platte is the main feature of interest.

The agricultural activity of this region is devoted very little to crop farming and more to hay production or grazing, due to the sandy texture of the soil and its low water-holding capacity. The light nature of the soil renders it very susceptible to blowing, hence it is necessary to keep a grass cover on it.

If the route happens to lie close to the Platte River, the traveller will have opportunity to see some of the irrigation canals which serve the flood plain and to note, as at Gothenburg, the way the ditches are carried by aqueducts over the streams and gullies which indent the valley sides.

The City of North Platte, Nebraska. At North Platte, the landing field lies a few miles east of the city on the bank of the North Platte River, practically at the point where the South Platte comes in from the south.

In the flight from Omaha westward the plane has climbed almost 2,000 feet, the elevation at North Platte being practically 3,000 feet above sea-level. North Platte, although it has a population of only 12,000, is the largest city of western Nebraska. It is distinctly a railroad-made town although with the recent development of irrigation it plays an important part in agricultural affairs. A state experimental farm is situated four miles to the south of the city. Sugar beets and hay constitute the chief farm products. Much of the hay is shipped westward to the Rocky Mountain markets.

Just to the southward of the city is the icing plant maintained by the railroad where over 80,000 cars of fruits and vegetables are iced each year.

North Platte occupies a unique position between the eastern and western centers of human activity for here the telephone and telegraph companies maintain their repeating plants and relay stations.

The North Platte and the South Platte Rivers. The journey west from North Platte lies between the North Platte and the South Platte Rivers. The North Platte River, in contrast with the South Platte, has a single well-defined channel, usually filled with water. The South Platte (Fig. 48) has a remarkably braided channel with an intricate interlacing of small streams. This difference appears to be due to the fact that the North Platte drains a part of the Rocky Mountains where the run-off is greater and where the melting snow maintains a more constant flow. The South Platte has a more meager supply. It has therefore developed a valley with a flood plain whose gradient is distinctly greater than that of the North Platte, this being a characteristic of all heavily loaded streams. During the dry period of the year the South Platte is entirely dry as all the water is removed for irrigation. A considerable flow, however, may be found beneath the surface of the flood plain. This accounts for the numerous trees to be seen in patches on the flood plain of the

FIG. 48. The South Platte River in central Nebraska showing its interlacing network of channels, and the many tree-covered islands.

South Platte whereas along the North Platte trees are almost absent because most of the water of that stream is confined to a single channel.

South of the river may be seen the prominent bluffs of loess (Fig. 49), rising abruptly 400 feet above the bottom lands. The loess is about 350 feet thick and lies upon the Ogalalla formation. This formation consists mainly of sand and gravel,

FIG. 49. The Platte Valley in central Nebraska. In the distance are the bare bluffs which rise to the level of the High Plains. Only along the course of the river are trees abundant.

cemented in some places by carbonate of lime into a resistant conglomerate, known as the "mortar beds." It becomes more and more prominent toward the west.

Just west of North Platte the flood plain of the South Platte River reveals scrolls of the old river channel now grown over in meadows. Southward there stretch vast tracts of cultivated ground. This is the region of dry farming, the rainfall being less than 20 inches per year. Much of the land is left in pasture. It is part of the central hay and live stock area. Many cattle are to be seen. The ranches comprise hundreds of acres of land, occasionally 1,000 or more. Often the most extensive view includes only one or two sets of buildings, at remote distances from each other. The farmstead is just a group of buildings, dominated by the windmill, and surrounded with a sparse assemblage of trees. Elsewhere not a single tree appears on the plain. The bluffs overlooking the Platte, however, are deeply gullied and tree-covered.

To the north between the North and South Platte Rivers is a rich belt of cultivated fields. Much of this is alfalfa, a crop which is commonly grown on the Platte and Loup River flood plains.

FIG. 50. The High Plains of western Nebraska. There are practically no streams over vast areas. The roads are straight and level for many miles; the farms are very large and make use of dry farming methods.

Far toward the north is the Sand Hill district, too far away to be easily seen, yet differing from the region to the south in showing many white patches, occasional lakes, and very little cultivation. The long peninsula of upland lying between North and South Platte Rivers is interesting because of the peculiar erosion which the loess soil has suffered. At Brule (Map No. 22), the route crosses into this region. Many "draws" or dry stream channels descend from the upland. They look like streams of mud. It was up one of these draws near Brule that the old California trail crossed over to the North Platte and it may still be seen. On the uplands the farms are very regular. The fields are large (Fig. 50), one quarter square mile in area being ploughed at one time. Every farm has its windmill, and here and there small ponds are to be seen. No fences and no trees interrupt the continuity of the fields and grasslands.

At Chappell the air route crosses the valley of Lodgepole Creek, a small strongly meandering stream, which it follows to Cheyenne.

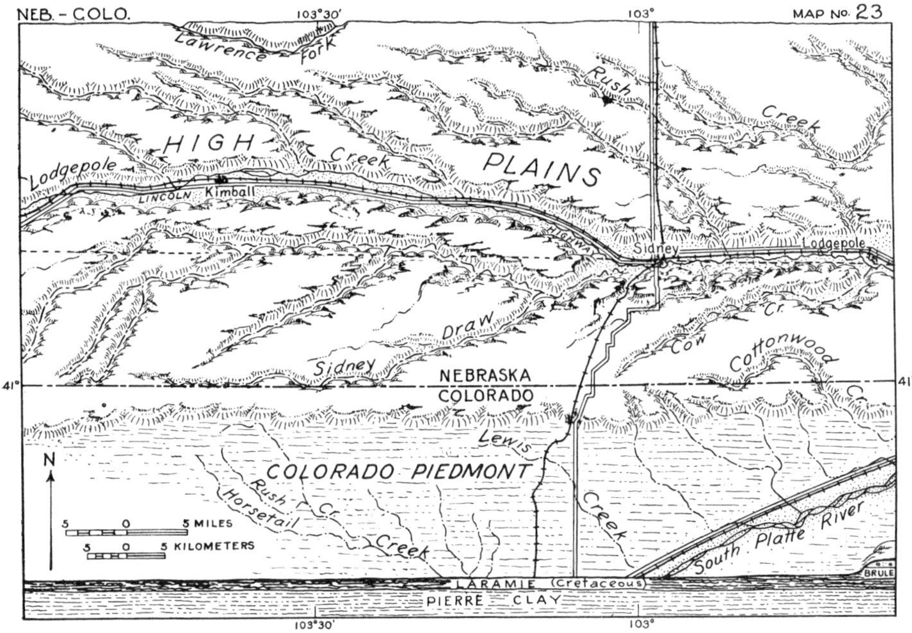

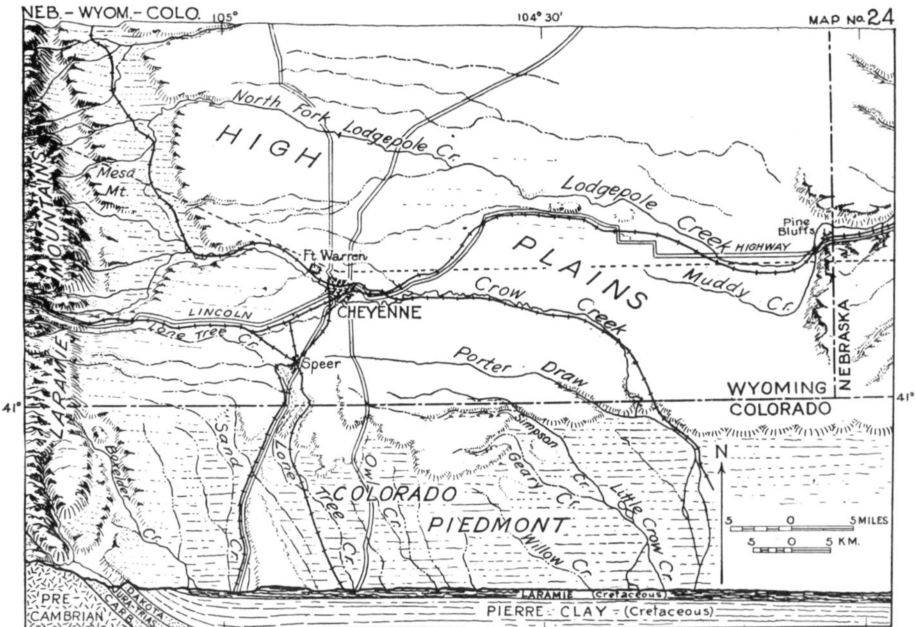

Route Maps No. 23 and 24

Route Maps No. 23 and 24

LODGEPOLE TO CHEYENNE

The Picturesque Mortar Beds. Beyond Lodgepole (Map No. 23) the white bluffs of the mortar beds become more and more conspicuous, owing to the disappearance of the loess cover. Benches and terraces along the valley walls, with here and there detached parts of the upland forming mesas and buttes and other architectural-like features, suggest the ruins of ancient cities. The Burlington Railroad going into Colorado passes between two such tablelands, just south of Sidney.[41]

At Pine Bluffs (Map No. 24) the air route enters the state of Wyoming. Here the country becomes extremely picturesque. The prominent bluffs of the mortar beds rise 200 feet or more above the river level. On these bluffs grow a few stunted pine trees. A tree is so rare in these sun-parched plains that these pines in their park-like setting attract especial attention. The bluffs may be traced a long distance to the north and south, and mark the western edge of the Ogalalla formation. They face west and look down upon the High Plains and the piedmont areas adjacent to the Rocky Mountains.

From Lodgepole westward the surface of the upland is dotted with prairie dog holes, looking like small sand hills, or like a battlefield which has been under shell fire. The cultivated fields, however, have been spared this invasion. It is possible that the observer may detect one of the prairie dogs if it happens to be in motion and if the plane is flying low.

The High Plains. From Pine Bluffs to Cheyenne the route passes over the Great Plains of eastern Wyoming. It comprises part of the High Plains, known in this section as the Cheyenne Tableland (Fig. 51). Its elevation is more than 5,000 feet above the sea. Although the land appears level, it actually slopes eastward about 20 feet to the mile, which is less than one-fourth of a degree and hence quite imperceptible, even to a person standing on the ground.

The Arikaree formation underlies the Ogalalla formation and extends west to the Rocky Mountains. It consists mainly of a loosely cemented sandstone. In southeastern Wyoming where the Arikaree formation occurs the High Plains show more diversity than in the Nebraska portion capped by the Ogalalla. This is due to the fact that the Arikaree has suffered considerable erosion and the original surface of deposition does not constitute the present surface of the plains as it does in the High Plains region capped by the Ogalalla beds. Mesas and buttes do not rise above the level of the Nebraska plains but they do appear in southeastern Wyoming.

The Agriculture of Western Nebraska and Eastern Wyoming. This section is devoted largely to wheat growing. Barley is also an important crop. The farms on the upland are very large (Fig. 50), one-fifth of them being 1,000 acres or over. The use of big farm machinery is therefore almost universal. This is favored by the level and unobstructed topography. Nearly all the agriculture is dry farming. Crops are grown without irrigation and with only a moderate rainfall, making special cultural methods necessary and to some extent limiting the kind of crops. Unplowed grazing lands still exist in the rougher areas and in some of the smooth parts that are still held by the stockmen of early days. There are a few irrigated ranches along Lodgepole Creek and Crow Creek. On these irrigated ranches wild hay and alfalfa hay are the principal crops. These hay fields occur in narrow ribbons along the edge of the streams. The stock ranches often comprise several

[41] The Ogalalla formation which produces the mortar bed topography is made up of gravel, sand and clay, each of which varies greatly in thickness from place to place. The resulting cliffs are therefore not so regular as in mesas carved out of more definitely bedded strata.

thousand acres each. The principal production of these ranches is beef cattle, commonly of the Hereford breed. There are, however, a few ranches that produce sheep and some that have a large number of milk cows. The unirrigated "dry" farms produce corn, potatoes, winter wheat, and barley. This dry farm development is less than twenty-five years old, and the farm buildings therefore are not so large and substantial as they are farther east. Some of them are not much better than the original homestead shacks. Windmills are rather plentiful, but they do not have the high towers that are common in Illinois. More and more the windmills are being replaced by gas engines. There are a few silos but they are not characteristic of the landscape. Many of the farmsteads are practically without trees.

Before reaching Cheyenne the route crosses the headwaters of Crow Creek (Map No. 24) which flows southward into the Colorado Piedmont. For a few minutes the traveller can look down into this piedmont region south of the air route, and note the striking contrast which it presents to the table-land above. It is everywhere well cultivated and appears to be gently rolling or actually flat. It is undissected. There are some meandering dry water courses, with red silt, this being the first intimation one has of the proximity of the "Red Beds" which outcrop along the Rocky Mountain front.

FIG. 51. Part of the city of Cheyenne, Wyoming, showing the extremely flat aspect of this part of the High Plains, whose elevation here is 6,000 feet above sea level.

Six miles east of Cheyenne and about one-half mile south of the Union Pacific main line track is the Cheyenne Field Station, operated cooperatively by the United States Department of Agriculture and the University of Wyoming. It has a small barn and two silos. There is a planting of trees around the farmstead and two rows of nearly a half-mile each border the north and east edges of a large field that is checkerboarded with experimental grain plots. These trees should constitute an easily distinguishable landmark to the air passenger as the air route passes practically over this station.

South of the Field Station is the ranch of the Wyoming Hereford Corporation, which is noted for breeding fine Hereford cattle and high producing Holstein cattle.

The Rocky Mountains. The approach toward Cheyenne over the Wyoming plains gives an opportunity to see the magnificent panorama of the Rockies rising gradually into view above the horizon of the plains. Almost 100 miles to the southwest the snowy summit of Longs Peak may be seen on a clear day. North of it extends the crest of the Front Range rising above the even surface of the Rocky Mountain Upland.

Straight ahead to the west the low Laramie Range is easily seen. Its unimposing appearance is apt to be a disappointment to the traveller who sees it for the first time and learns that this is where the air route crosses the Rockies and reaches its highest elevation above the sea. The low sag through which the Union Pacific and the Lincoln Highway go lies immediately head. The air route, however crosses the range farther north.

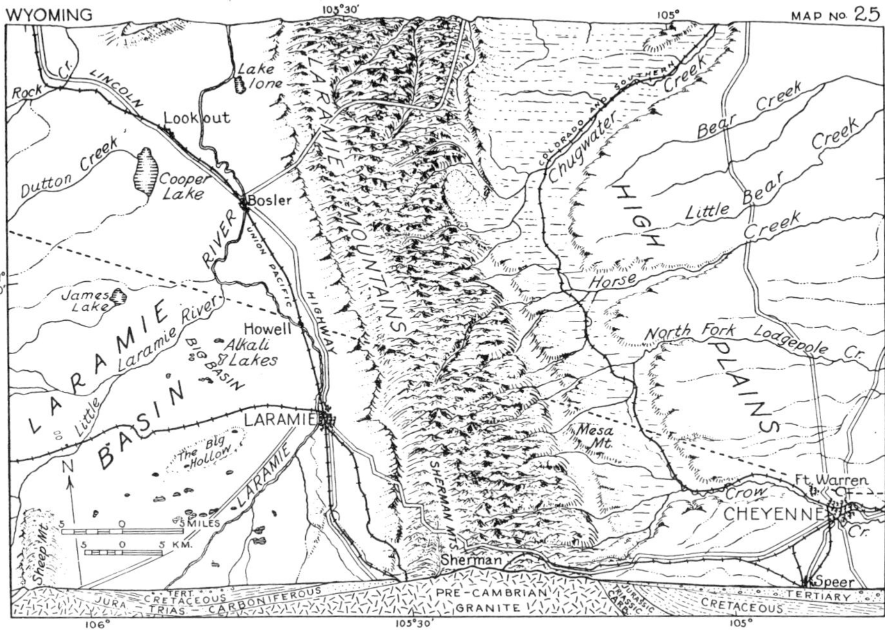

Route Map No. 25

CHEYENNE TO LARAMIE

The Rocky Mountains. Immediately after the take-off at Cheyenne there is spread before the air traveller a far-flung view which encompasses much of the Southern Rockies. Longs Peak is visible, almost 100 miles away to the south, as well as Twin Peaks and other parts of the Front Range, all beautifully snow-capped and usually standing out sharp and clear in the bright western light. Straight to the west are the low Laramie Mountains, beyond which rises the serrate form of the Medicine Bow Range with its snowy peaks. Below and off to the east stretches the limitless expanse of the High Plains almost undiversified by any topographic form. Occasional ponds for irrigation and for the watering of live stock may be noted. The route passes almost over Fort Warren on the outskirts of Cheyenne and heads toward the northwest.

The Rocky Mountain Front. The most important view now is to the south. There the surface of the Great Plains may be seen rising with an even slope from an altitude of 6,000 feet at Cheyenne to over 7,000 feet where it overlaps the flank of the Laramie Range. This "land bridge"[42] is the only remnant of the High Plains which still touches the mountain mass. Almost everywhere along the front of the range the Tertiary formations constituting the plains area have been eroded so as to leave a lower tract of land between the mountains and the plains. This lowland, or piedmont belt, where crossed by the airway just northwest of Cheyenne, is quite narrow and serves as a natural route for the Colorado and Southern Railroad going

[42] This "land bridge" provides a relatively easy grade for the Union Pacific Railroad, as well as for the Lincoln Highway, over the Laramie Range.

It is sometimes called the "gang plank" connecting the mountains with the plains.

northward along the eastern base of the Laramie Range. As the front of the range is approached the upturned Cretaceous, Jurassic, Triassic, and Carboniferous beds come into view. These form a bold hogback[43] which is by no means continuous but is interrupted by many gaps. The sedimentary beds are upturned in some places to a nearly vertical position. In Mesa Mountain, however, just south of the air route, they are almost horizontal. Still farther south where the land bridge overlaps the mountains, they are buried. The Casper formation (Carboniferous) and the Chugwater formation (Triassic and Permian), each about 1,000 feet thick and consisting largely of red sandstone, account for the brilliant colors of the hogback ridges. These formations in turn are capped by the lighter-colored beds of Cretaceous age. Where the hogbacks are missing the front of the granite upland looks out boldly toward the plains.

The Laramie Range. In spite of the fact that this is the highest point on the air route between New York and San Francisco, the Laramie Range does not appear very grand or striking in appearance. This is due partly to the fact that the Great Plains to some extent rise up to its summit and also to the fact that it is not a range of rugged peaks. It is rather a flat-topped upland. Its surface is a peneplain.[44] Views in all directions reveal its fairly level top. Some higher irregular masses dominate the upland level and the surface is made interesting by the presence of many peculiar pinnacles, buttes, knobs, castles, and other architectural-like forms, due to the weathering of the jointed granite which makes up to the core of the range. The jointing is so regular in places as to give the effect of sedimentary beds. The streams flowing toward the east have in most cases cut deep gorges into the granite. On the upland surface the stream courses can easily be traced by the scattered growths of evergreens. This gives a park-like aspect to the mountain area. Several secondary roads cross over it, apparently with little difficulty. It is, according to Professor Davis "just what a peneplain should look like," by no means flat, but with extensive areas of rolling land.

In this upland area lies the Pole Mountain Forest and Military Reservation which furnishes summer grazing for a large number of live stock, principally sheep. This region, together with the adjacent foothills, is part of the range land of the Warren Livestock Company, one of the largest sheep ranches in the United States. This company was founded by Francis E. Warren, for many years the senior senator from Wyoming. Fort Francis E. Warren, the military post on the edge of Cheyenne, was named for him. The holdings of the Warren Livestock Company extend from a line a little to the east of Cheyenne westward to within five or six miles of Laramie. South of Cheyenne, starting almost at the edge of the town, the company owns land extending twenty or thirty miles south over the border into Colorado, and starting six or eight miles north of Cheyenne has another tract that extends probably forty miles north into the region near Bear Creek. The Warren Livestock Company carries about 50,000 sheep during the winter. In the summer before the lambs are sold the total number of sheep on the Warren range probably exceeds 75,000 head. It can be well understood why one of Senator Warren's colleagues dubbed him "the greatest shepherd since Abraham."

The regular route of the airplane crosses the Laramie River about eight miles to the north of the town of Laramie.

43 A hogback is a ridge formed by a tilted resistant formation which projects above the general surface of the country. Hogbacks occur in many parts of the western United States. They are especially striking along the front of the Rockies in Colorado.

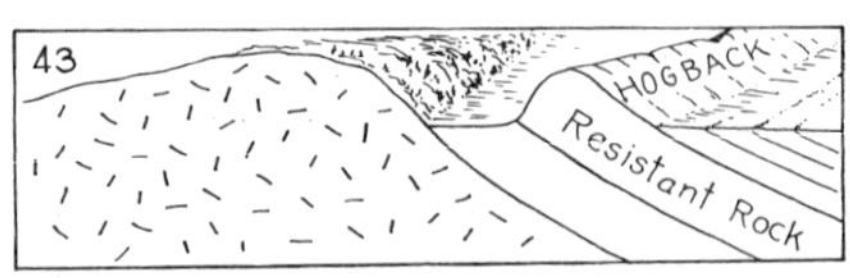

44 A peneplain is a flat expanse of country which at one time was worn to sea-level. Most peneplains have been elevated above sea-level and now constitute uplands, with valleys cut below the upland surface. See note 3.

On the western side of the range the deep vermilion-colored rocks of Carboniferous age again make their appearance, this time forming an escarpment of marked continuity. They are a great deal more colorful than the somewhat sombre red granite of the upland tract. The hogbacks are not ridge-like, but, due to the very gentle dip, have the form of a tilted plateau with a serrate, eastern-facing escarpment.

The Laramie Basin. Attention may now be directed to the broad expanse of the Laramie Basin with its scattered alkali lakes and white salt flats (Fig. 52). Wind-scoured hollows, not easily detected from the air, constitute the chief topographic feature of the plain. Laramie is easily within view. It is situated close to the western flank of the Laramie Range on the Laramie River. This stream, strongly meandering and abundantly supplied with water from the snows of the Medicine Bow Range, serves to irrigate a zone which stands out rich and green in contrast with the barren lands on all sides. Snow fences along the railroad, in some places three abreast, indicate that this is a region of considerable snow fall and probably strong winter winds. To the northeast the northern tip of the Laramie Range may be seen gradually

FIG. 52. The arid expanse of the Laramie Basin, almost devoid of vegetation. An alkali lake in the foreground with salt-encrusted.plains.

disappearing beneath the plains. Off toward the west is the Medicine Bow Range with its level summit profile strikingly displayed. It is surmounted by a group of serrate snow-capped peaks.

The average altitude of the Laramie Plains is about 7,000 feet with the grassy foothills rising 1,500 feet higher. This great altitude causes cool nights and short seasons that limit the types of agriculture. West of Laramie the rainfall is so small that little or no successful dry farming is done along the air route. The principal crops grown on the irrigated farms are wild hay, locally called native hay, and alfalfa hay. Small grains such as spring wheat, oats, and barley, potatoes and all root crops are successfully grown, and occasionally sunflowers are grown for silage. In places there are rather large tracts of irrigated land where the production of giant potatoes and alfalfa hay is carried on. At the edges of these tracts and in the higher mountain valleys and along narrow streams there are isolated stock ranches that produce hay that is fed in the winter to cattle and sheep which graze in the mountains and on the unirrigated plains and foothills during the greater part of the year.

The Laramie River, the Lincoln Highway and the double-tracked Union Pacific Railroad run parallel for several miles north of Laramie. The air route crosses the railroad at a section house and water tank called Howell, which is also the place where the railroad crosses the river.

Due east of this point is a group of low buildings with a windmill which are the ranch headquarters of the King Brothers Company. This ranch is not so large as that of the Warren Livestock Company, but it generally carries between 15,000 and 20,000 sheep during the winter and has about 200 square miles of land extending over the Laramie Mountains from the Union Pacific on the west to the Colorado and Southern on the east.

At Bosler, to the north of the Union Pacific tracks, there is a group of irrigated farms and along the Laramie River north of Bosler are a number of the large old-type cattle ranches that have survived from the early days.

Soon Cooper Lake, off toward the north, may be seen surrounded with white incrustations of sodium carbonate. Near at hand, to the south is James Lake, which obviously has no outlet and because of its many shore lines appears to have varied greatly in extent from time to time.

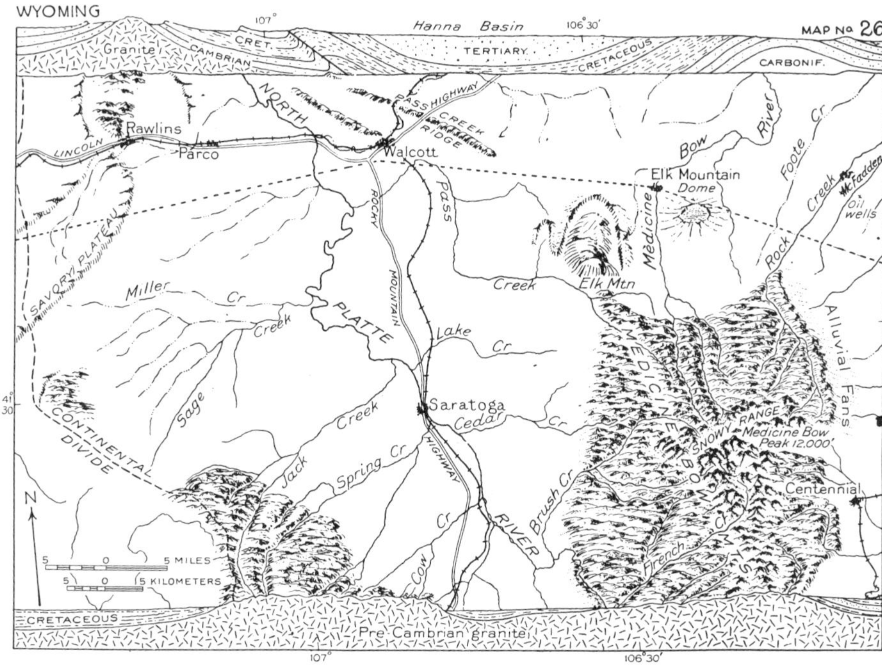

Route Map No. 26

ELK MOUNTAIN AND RAWLINS REGION

The Medicine Bow Mountains. On the western side of the Laramie Basin rises the barrier of the Medicine Bow Mountains, around which the air route must pass. The white peaks of the "Snowy Range," with an elevation of over 12,000 feet, and standing like sentinels on the Medicine Bow upland, dominate the view. Long sloping alluvial fans of recent gravels extend eastward from their flanks. Below these gravel plains the streams have incised themselves in broad flat-floored valleys, or arroyos,[45] for the stream beds seem to be dry. Nevertheless, numerous cattle may be seen along the bottom lands, taking advantage of the better grazing due to the moisture which there comes to the surface.

In the vicinity of Elk Mountain the ranches are more scattered. In most cases they are true stock ranches that are merely a headquarters from which to run sheep and cattle on the range. The sheep are generally on the range throughout the year. In the summer they go into the high grassy meadows and treeless slopes of the Medicine Bow and Sierra Madre Mountains, at the upper edge of the timberline. In the winter they are on the sagebrush-covered flats where there is very little snowfall. In the Wyoming climate nearly all of the cattle have to be fed considerable hay in the winter time; consequently the cattle ranches all have hay meadows of varying sizes. There are groups of these small cattle ranches around the foot of Elk Mountain and other mountains of the Medicine Bow region. On these high-altitude cattle

[45] Most of the mountains in the arid lands of the western United States have long gently sloping deposits of sand and gravel extending from the mouths of the canyons onto the surrounding plains. The streams which deposited these "fans" are now, in times of flood, cutting channels below their surfaces.

ranches that are found in the narrow valleys of the mountains the buildings are often of logs from the near-by timber, some even having dirt roofs. Corrals are constructed of the light, straight poles of the lodgepole pine, and in some places there are miles of "buck" fence constructed from light poles, surviving from the early days. The cattle on these ranches are usually of the Hereford breed, and the ranches vary in size, the number of cattle being controlled largely by the amount of hay-land that is available. It is a common rule among these ranchmen to allow a half a ton to a ton of hay for each head of cattle that they expect to carry through the winter. Sometimes in mild winters all the hay is not needed and in the following spring and summer the mountain meadows will be dotted by haystacks left from the previous summer.

The Oil Domes of Southern Wyoming. Just before the northern end of the Medicine Bow Range is crossed, a number of oil wells may be noted near the town of McFadden, which lies slightly to the north of the air route, and almost on the axis of one of the domes so numerous in this part of Wyoming. Teapot Dome, north of

Fig. 53. Hogback formed by the eastward-dipping "Red Beds," on the eastern flank of Elk Mountain about 50 miles west of Laramie, Wyoming. East is to the right in the picture.

this region is a famous example. Parco, a few miles farther west, on the air route, derives its oil from a similar structure.[46]

The railroad and highway lie many miles to the north and out of sight. The lighted air route follows them closely and is used in bad weather, but ordinarily planes take a direct path, and depart from the lighted way by a score of miles or more.

Much of this region has been strongly warped and tilted. The arched structure of one of the smaller domes may be clearly seen east of Milo just before Elk Mountain is reached. Great hogbacks flank the north side of Elk Mountain (Fig. 53) which the air route usually skirts closely. East bound planes flying straight through from Salt Lake City to Cheyenne without stopping at Rock Springs take a more southerly

[46] The structure of this dome is clearly shown in the geological section at the top of Map 26. In the Parco, or Rawlins, dome the granite core is exposed in the central part of the dome where it forms low hills. Surrounding this dome is an almost complete circular rim or hogback.

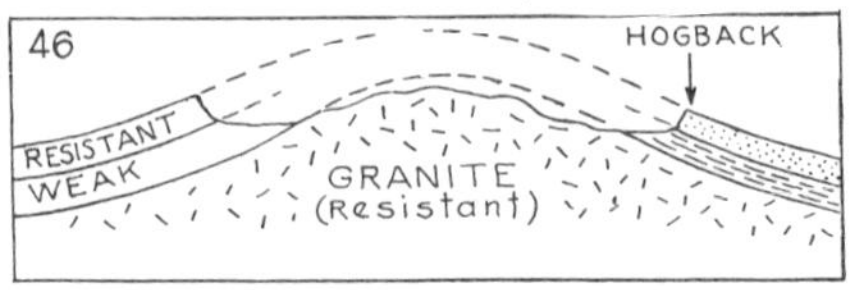

route and go between Elk Mountain and Snowy Range. Passengers on such planes therefore may not see some of the features herein described. Beyond Elk Mountain to the west is the broad basin of the North Platte River. Across its floor may be seen the gleaming white ribbon which marks the Rocky Mountain Highway running southward from Walcott into North Park. From Walcott westward the route practically follows the Union Pacific Railroad and the Lincoln Highway to Rock Springs.

South of Walcott there is a large area of land, at present covered by sagebrush, that constitutes the proposed Saratoga Irrigation Project. The Bureau of Reclamation of the United States Government has under consideration a plan to irrigate about 60,000 acres of this flat from the North Platte River. It lies on the east side of the river and extends northward toward Elk Mountain as far as Pass Creek, and south to Saratoga. If this land is irrigated it will be divided up into farms of 80 to 160 acres, and the production of grain, alfalfa and potatoes will be the principal types of farming.

Parco, on the eastern flank of the Rawlins uplift, is the most active oil producing town along the air route. The name of the town can easily be read on the roof of one of the buildings but even without this the town can be identified by its numerous oil tanks.

North of Rawlins about 30 miles, may be seen two mountain groups, the Seminoe Mountains, and the Ferris Mountains, lying just west of them. The Ferris Range may be distinguished by the striking white scallops on its southern flank due to vertical beds of limestone which have resisted erosion.

The dome on which Rawlins and Parco are situated is much more strongly uparched than are most of the domes in this part of the state.[47] This explains the presence of granite in its middle part. Savory Plateau, made up of westerly dipping beds, is a continuation of the hogback which rims the western side of this dome.

[47] Oil wells are situated, not with regard for the topography but in relation to the rock structure. Oil domes may be high or low topographically, depending on the amount of erosion which has taken place.

The Parco oil field is situated not in the center of the dome where the granite outcrops but at the southern end where the structure is like that shown in the accompanying sketch.

47

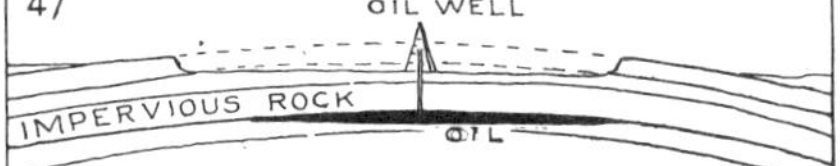

A Low Dome

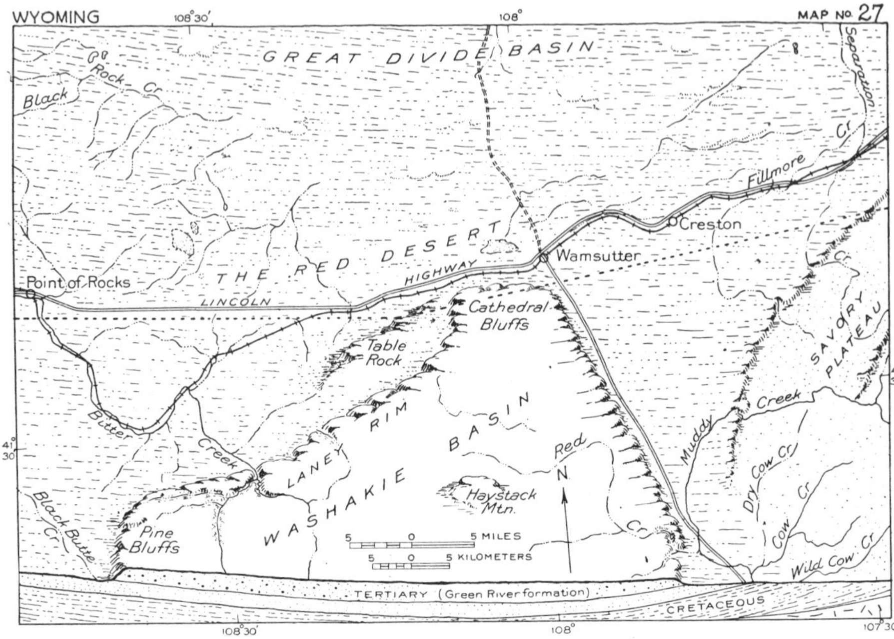

Route Map No. 27

THE GREAT DIVIDE BASIN

The Great Divide Basin. After passing Rawlins (Map No. 26), the route follows a straight course over a barren tract of desert (Fig. 54) known as the Great Divide Basin. This is called a basin because from all sides the rocks dip very gently toward the center where they are virtually horizontal.[48] It is part of the Continental Divide separating the waters of the Atlantic from those of the Pacific, that is the headwaters of the Missouri system from those of the Green River, which is tributary to the Colorado. Unlike most of the Continental Divide which follows mountain crests, this part is extremely flat. In fact there are very few actual streams in the area. Dry water courses or arroyos lead to shallow depressions containing alkali lakes. These, during the summer, evaporate and the plain becomes covered with a white incrustation of salt. One of these lake beds lies a few miles west of Wamsutter, and several more are visible to the north before Point of Rocks is passed. The margin of the Great Divide Basin is marked by prominent outward-facing escarpments, where the more resistant beds outcrop. Several of these are crossed shortly after passing Rawlins, the largest constituting the edge of the Savory Plateau.

Just beyond Wamsutter the route crosses over the end of a high plateau or table-land which seems to rise above the level floor of Great Divide Basin. This plateau, known as Washakie Basin, has been carved by erosion out of the Tertiary[49] rocks which

[48] In structure the Great Divide Basin is a broad syncline extending from the Rawlins uplift on the east to the Baxter Dome at Rock Springs on the west. The geological section in the front of Route Map No. 27 shows how the beds dip or slope toward the center. The whole region resembles a pile of flat plates with the smallest one on top. The edges of the plates form the outward-facing escarpments, which are conspicuous features on both the eastern and western sides of the region.

[49] The Tertiary beds of central Wyoming are vast deposits of gravel, sand, and clay known as the Wasatch Group, which were laid down by streams in basin areas between the different ranges and mountains of the Rocky Mountain System. Erosion since that time has removed a large proportion of the original deposit, the remnants of which now constitute plateau areas such as Washakie Basin.

form the uppermost strata in this part of Wyoming. It represents merely the remnant of the formations that originally extended all over this region, covering the Cretaceous beds now exposed at the surface in Great Divide Basin.

Picturesque and architectural in their details, the cliffs, descending from the margin of this upland, present a magnificent sight to the air traveller because he is enabled to observe them so closely.

The red wall of Cathedral Bluffs, constituting the northern end of what might well be called the Washakie Plateau, continues southward to form Laney Rim which may be seen extending for many miles, far away toward the southwest to Pine Bluffs. Then, just to the west of Cathedral Bluffs is an outlying mesa, called Table Rock.

The Red Desert. The central part of Great Divide Basin is called the Red Desert. As the name suggests, the dominant colors are red, russet, brick-red, and vermilion, but there is every tone of gray and brown, with shades of green, purple, and yellow. Unlike the colors of an eastern landscape, those of the Red Desert are not dependent on the season, for there is little vegetation to hide the coloring of the rocks and soil.

FIG. 54. The Great Divide Basin, Wyoming, showing the mesas which bound it on the south.

Blackish bands of sombre hue interrupt the display of brighter colors. These are seams of coal, or rather lignite, which are included in beds of Cretaceous age throughout the whole Rocky Mountain area.[50] Between Creston and Wamsutter, the black soil is certain to attract attention.

To ride over so vast an unsettled area and see not a sign of human occupation, except the little ribbons of railroad and highway for a hundred miles or more is impressive indeed. Throughout this distance there is only one road which crosses the desert. It runs north from Wamsutter. Hardly more than a trail, it is nevertheless certain to catch the eye, as it winds its way through scattered bunches and tangled growths of stunted sagebrush and greasewood.

Despite the sparsity of vegetable growth, the Red Desert is a winter sheep range, and great flocks of sheep are likely to be seen here at that time. In the summer the sheep are herded in the mountains and the desert appears to be an uninhabited wilderness.

[50] Lignite occurs widely in Montana, Wyoming and Colorado and forms the chief fuel of the mountain states. It is of great local importance but can not stand shipment where it is exposed to the weather very long, as it sloughs and disintegrates rapidly.

But the severity of winter at higher altitudes forces the herders to come here with their canvas-covered wagons and set up camp on sheltered hillsides. In these protected spots are located the bedgrounds where the sheep stay at night. On sunny days as the shadow of the plane races over the ground there is a great commotion among the sheep although the men on duty below welcome each day this reminder of the outside world from which they are for so long detached.

As Point of Rocks is approached the eastward dip of the formations becomes noticeable. This dip is due to the strong tilting of the beds on the eastern flank of the Baxter Dome, next to be described.

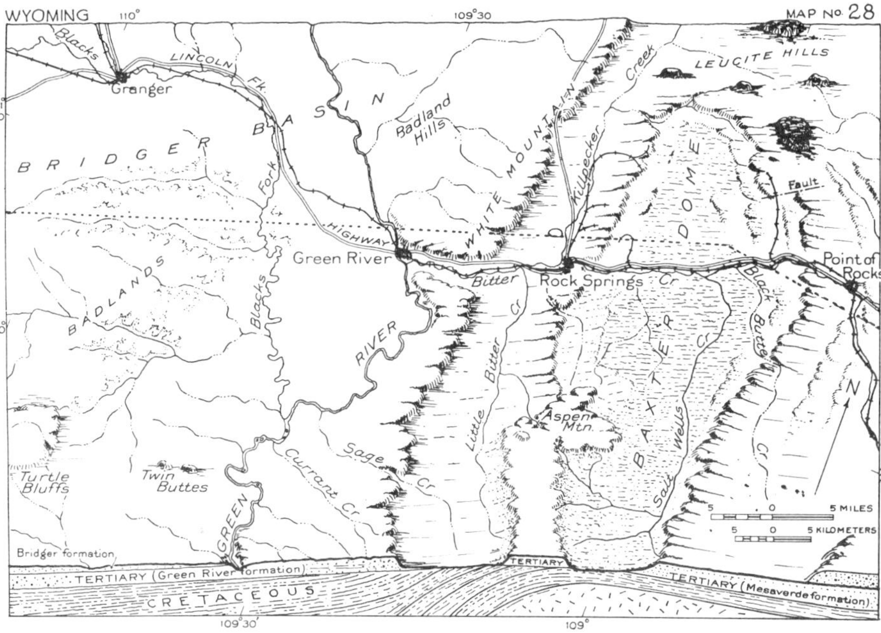

Route Map No. 28

ROCK SPRINGS–GREEN RIVER REGION

The Baxter Dome. Between Point of Rocks and Green River the air route traverses one of the most interesting geological structures encountered on the transcontinental trip, the Baxter Dome. It is interesting, not because of any very spectacular scenery but because it is an example of geological structure which clearly shows the effect of rock folding upon topographic forms. The Baxter Dome is only one of many domes and uplifts which have bowed upon the rocks of Wyoming. Mention has already been made of several others. Some of them are small, being only a few miles across, but the Baxter Uplift is ninety miles long from north to south and fifty miles wide. Most of the uplifts are gentle ones, that is, they do not form high mountains, although in other parts of Wyoming there are some big arches, like the Bighorns, the Wind River Range, and the Black Hills. The Baxter Uplift, unlike the higher domes, is remarkably simple. The rocks have been bowed up so that they dip away in all directions from the central portion. Erosion has removed the uppermost beds from the crest of the arch and has exposed there the lower or underlying strata. Because of their weakness these lower beds have been extensively worn away with the result that the central portion of the Baxter Dome is now a topographic basin. It is still, however, a dome in structure. The margins of the eroded formations constitute escarpments facing toward the center of the uplift. These encircle the basin of erosion almost without interruption, and are formed by the edges of tilted strata of hard resistant sandstones of the Mesaverde formation. The circular pattern of these ridges is suggestive of the layers of an onion which has been cut through the middle. Some of the innermost layers contain valuable coal beds which make this one of the most important fields of the western United States. The air route, between Point of Rocks and Rock Springs, passes over the northern end of the dome and thus the rimming escarpments at that end may readily be observed, but the southern end is too far away to be clearly seen.

Like many of the other domes, the Baxter Dome has been the locus of oil and gas accumulation (See Note 47). Near the center of the area several gas wells may be seen from the plane but most of them are in the southern part of the basin. The pipe line which carries this gas to Salt Lake City may later on be noted in many places, west of Green River, where it looks like a ploughed furrow along the valley sides.

The basin country, occupying the central part of the dome is barren and desolate. From the air the general appearance is that of a vast, broad valley, whose surface is nearly level. But closer observation shows that it is traversed by numerous smaller valleys and dry gulches, with a relief in places from 100 to 300 feet. Much of the area is cut into sharp ridges and benches, but the sombre color of the treeless hills causes each ridge to merge with the next, forming a broad rolling expanse whose prevailing tones are drab and gray.

Several miles south of the railroad a broad mesa may be detected apparently occupying the middle of the Baxter basin. This is Aspen Mountain. It consists of the Bishop conglomerate, one of the later Tertiary gravel formations which lies in a horizontal position and entirely conceals the tilted beds underneath.

The Leucite Hills. Ten to twenty miles north of the air route, on the margin of the Baxter Uplift, a number of dark buttes and mesas may be descried. These are the Leucite Hills, which consist of igneous rocks of volcanic origin, the lava having worked its way upward into the sedimentary rocks at the time of the warping of the beds to form the dome. Volcanic necks, sheets, and dikes have been exposed by the later erosion of the sedimentary rocks, but there are also more recent volcanic cones and surface flows.[51] Still farther away, on the northern skyline may be seen the snowy peaks of the Wind River Range of central Wyoming.

Rock Springs. The airport at Rock Springs is situated on the western flank of the dome at the foot of the White Mountain escarpment. This ridge, on the western side of the Baxter Uplift, corresponds with Pine Ridge and Laney Rim on the eastern side and is formed by the Green River beds of Tertiary age. White Mountain receives its name from the light-colored sandstones and shales which dip quite strongly westward and descend to river level at the town of Green River. The aridity of the Rock Springs region may be appreciated from the fact that the water supply for this small town has to be pumped from the Green River 15 miles away.

The Green River Formation. When the airplane leaves the airport, going west, it follows the valley of Bitter Creek and flies close to the bluffs of the Green River series. These beds are made up of a countless number of very thin sandy and shaly layers so that the cliffs have a minutely banded appearance. Although these rocks give no outward sign of the presence of oil they yield on distillation as much as 31 gallons to the ton. Between Rock Springs and Green River the plane goes over the lower bench of the Wasatch formation which underlies the Green River beds and forms a pronounced terrace.

[51] The name "Leucite" comes from the mineral leucite, which is common in the volcanic rocks of these hills. A "volcanic neck" is the core of a former volanco which has been destroyed by erosion. A "sheet" is a horizontal layer of lava which has been intruded between sedimentary beds. A "dike" is a vertical layer of lava which has been intruded into a crack or joint. Volcanic "cones" and "flows" are surface features.

In the accompanying sketch the original cone and lava flow have been completely destroyed. The remnant of the "sheet" forms a mesa. The dike forms a wall.

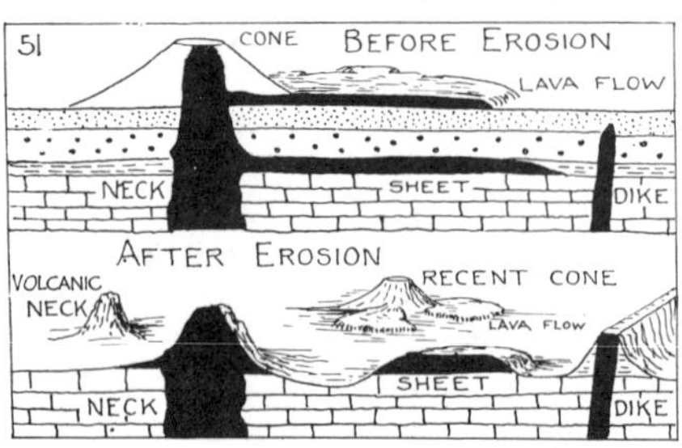

At the town of Green River the upper part of the Green River formation, called the Tower sandstone, is carved into many curious and picturesque forms, natural monuments and castle-like structures which are easily seen from the flying plane passing close to them.

The Bridger Basin. The route west from Green River lies over the center of the Bridger Basin with its almost horizontal formations. The region, however, is beautifully dissected into characteristic badlands. Between the streams there are broad flat tracts of upland supporting scattered bunches of sagebrush but the valley sides everywhere present an amazing amount of detail.

To the south, 20 miles away, Twin Buttes and Turtle Bluffs are easily seen. These, like Aspen Mountain near Rock Springs, are remnants of the Bishop conglomerate overlying the Bridger formation in which the badlands are cut.

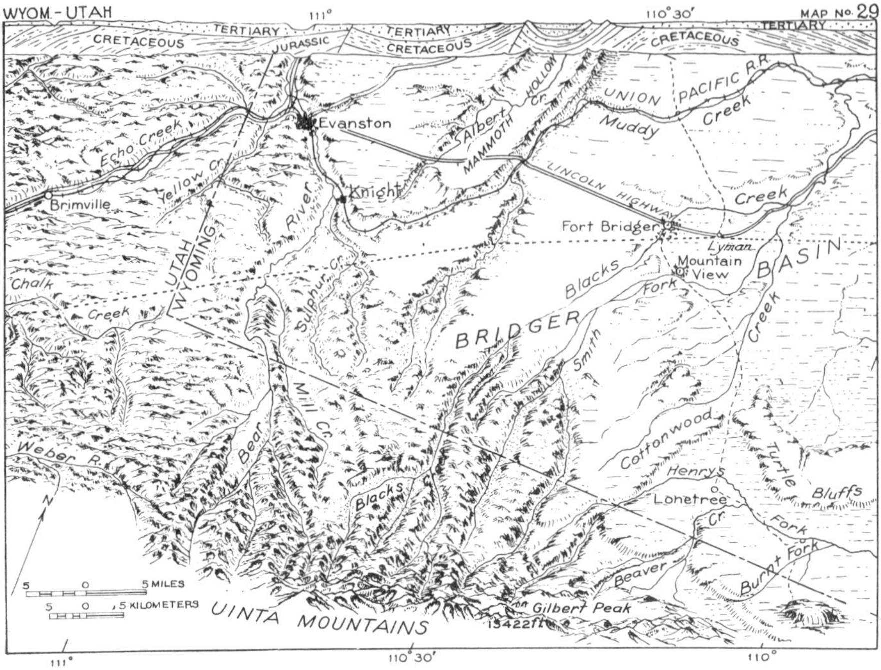

Route Map No. 29

THE FORT BRIDGER REGION

The Bridger Basin. In the southwestern corner of Wyoming, lying between the Baxter Dome on the east and the Wasatch Mountains on the west is the Bridger Basin. Like the Great Divide Basin this is also a broad shallow syncline, the formations dipping gently toward the center. Unlike the Great Divide Basin, however, the Bridger Basin is drained by many streams which have opened out broad flat valleys below the general level of the Tertiary beds forming the original floor of the basin. Physiographically, this country is said to be post-maturely dissected.[52] The air route passes alternately over the upland spurs with their desert-like appearance, and the rich green bottom lands which are strongly reminiscent of the "meadows" in Yellowstone Park and adjacent regions. In these broad valleys lie the towns of Lyman, Fort Bridger, and Mountain View. The aspect of these lowlands is utterly different from that of the upland which borders them. Irrigation canals are to be seen running throughout the bottom lands. Grain, alfalfa, and native hay are the chief farm products. Many of the houses are surrounded by trees, some of which are fruit-bearing. In recent years poultry and dairy industries have been started in the Bridger Basin.

On the southeast edge of Lyman is a state experiment farm, operated by the University of Wyoming. This farm can be distinguished by a large white dairy barn somewhat of the type found in northern Iowa. There are also two small silos.

[52] That means it has been greatly denuded by erosion so that hardly any of the original surface remains. The streams flow in broad valleys and lowlands, the narrow mesa-like divides between them being the only part of the original upland which remains.

52 BROAD FLAT-FLOORED VALLEYS IN THE BRIDGER BASIN

(*Photo by United Air Lines*)

Fig. 55. The Uinta Mountains, a view southward from the Bridger Basin of western Wyoming.

Immediately beyond the park-like valley of Blacks Creek, with its scattered clumps of pines, the plane climbs to go over Bridger Butte. This is a part of the Tertiary upland. Here, cattle and sheep may be seen although the pasturage on the upland seems to be very scanty. Everywhere there is an abundance of sage-brush. Close observation may detect the presence of prairie-dog holes. This region is important as a winter range for sheep. Off to the north the Lincoln Highway crosses the low tract of country which represents the level to which erosion has thus far reduced the area. Here and there along the route may be seen the pipe line which carries gas from the Baxter field to Salt Lake City. It looks like a road or furrow following the hillsides but going up and down over the minor irregularities.

The Uinta Mountains. During the journey from Rock Springs to Salt Lake City the Uinta Mountains are visible 30 to 50 miles away to the south (Fig. 55). The air route runs almost parallel with the range. On a clear day it is possible to distinguish some of the more salient features, especially near the western end where the route comes within 30 miles of the main crest. Many of the peaks of the Uintas rise to elevations of 13,000 to 14,000 feet. The crest of the range, however, is not strikingly serrate. It is rather an upland into which glacial cirques[53] have been cut. The remnants of the upland now constitute the higher peaks. Viewed from certain directions, however, the upland level is not very prominent and the summits seem to be sharp pointed.

As the traveller crosses this rugged country in southwestern Wyoming he may observe at several points the beacons marking the air route. These are situated on the tops of conspicuous peaks, and the plane usually flies not very far above them. This, indeed, is one of the most spectacular parts of the transcontinental journey.

53 GLACIAL CIRQUES

[53] A glacial cirque is a cup-shaped valley gouged out of a mountain by the action of a glacier. Practically all of the western mountain ranges have cirques on the sides of their higher peaks. In the accompanying sketch glaciation has just begun to destroy the mountain area. See also Note 56 on page 108.

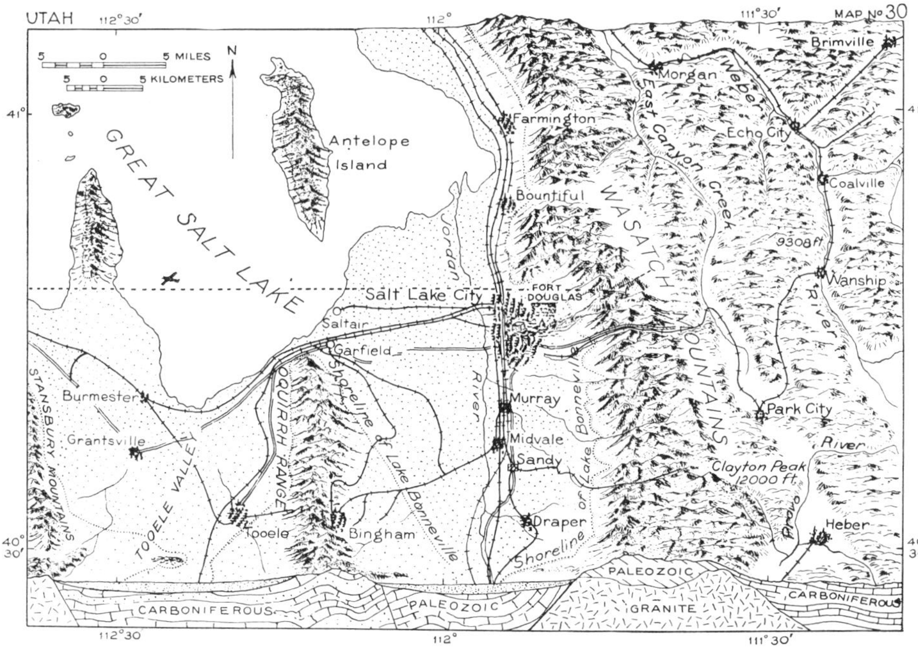

Route Map No. 30

THE SALT LAKE CITY REGION

The Wasatch Mountains. The eastern or back slope of the Wasatch Mountains, crossed between Coalville and Salt Lake City, presents some of the most rugged and savage scenery to be found in the United States (Figs. 56 and 57). The geologist calls this region a maturely dissected complex mountain area, which is another way of saying that it is a maze of canyons, deep narrow defiles, sharp crested ridges and peaks.

Weber River and several of its tributaries flow through this region in wide level-floored valleys (Fig. 58) and it is in these rich irrigated valleys that the people who have settled in these mountains must necessarily live. Two things have attracted people to this country, namely its mineral wealth and its agricultural possibilities.

Coalville, situated on the western margin of the Bridger Basin and in the foothills of the Wasatch, is one of the most important coal mining cities on this part of the air route. The mining camps in Weber Valley and in Grass Creek Valley just to the north are in full view. The coal occurs in beds of Cretaceous age, much disturbed by folding and faulting. West of Coalville, red and white Jurassic limestone with scattered clumps of pine trees add a touch of color to the landscape. The topography becomes extremely rugged but the dip of the geological formations and the structure of the rocks may easily be noted because of the scanty vegetation and soil cover. In some of the narrow valleys reservoirs for irrigation have been built. Several miles south of the air route, in the headwaters of East Canyon Creek, one of the broad flat valleys of the region, is Park City, another mining town. Its importance is due to metals rather than to coal. Since their discovery in 1869 the mines in this valley have yielded a total value of $200,000,000 in gold, lead, silver, copper and zinc.

The snow-capped peaks of the Wasatch next command attention, as the plane climbs steadily to the crest of the range. It is a splendid and inspiring sight. All

Fig. 56. The maturely dissected upland of the Wasatch Mountains in eastern Utah. Practically all of the forms are due to stream erosion. Occasional small incipient cirques may be detected.

signs of human occupation give way to a sea of tumbling peaks and ridges and wild mountain scenery (Frontispiece). The forest cover, mainly spruce, becomes heavy because of the more abundant rain and snow at these higher altitudes. On the steep walls of the deep ravines and gorges the red beds, making up the mountains, outcrop between the trees.

Through a narrow pass at the summit of the Wasatch Range the route makes its way to emerge upon its western front with the magnificent and almost infinite panorama of the Great Basin spread out ahead. The scenic features now are all too numerous to be seen in the few brief moments which remain before the landing field is reached at Salt Lake City. The distant view encompasses Great Salt Lake and the many ranges rising up like islands above the vast level plain of the Great Basin.

Fig. 57. The western face of the Wasatch Range, a view looking south from a plane which has just left the Salt Lake City Airport. The Bonneville beach near the foot of the range may be seen, and in the foreground extensive and only slightly dissected alluvial fans.

FIG. 58. The back slope of the Wasatch Mountains, eastern Utah, descending to the flat-floored valley of the Weber River near Coalville.

But near at hand there are more important things. The physiographer should immediately rivet his attention upon the steep western front of the Wasatch Range. This range is famed in geological literature as an example of a block mountain formed by uplift along a fault plane.[54] Usually mountains of fault origin have very straight base lines and, if not reduced by erosion to an old condition, they rise abruptly above the plains. The base of the Wasatch Range (Fig. 57, and also Fig. 2) is far less straight than might have been anticipated. The explanation is that erosion has worn

FIG. 59. Front of the Wasatch Range, north of Salt Lake City, showing the uppermost or Bonneville terrace formed by the waves of Lake Bonneville during its highest stage.

[54] Most of the Basin ranges are block mountains, so-called because they are blocks of the earth's crust which have become dislocated and usually tilted. The front face is therefore steeper than the back slope.

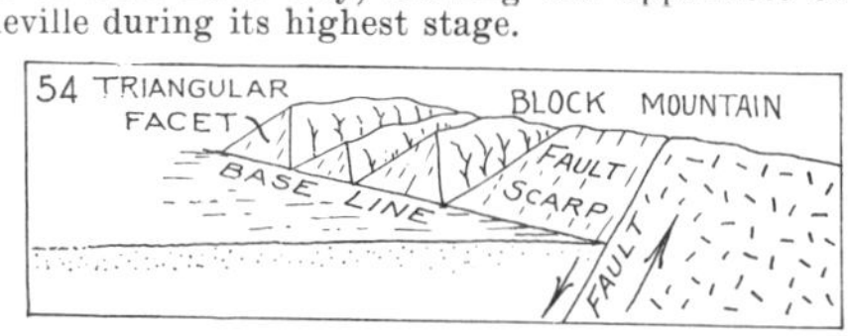

the front of the range back of the original fault line in an irregular manner. Moreover, the "triangular facets" which are such conspicuous features on the front of the Spanish Wasatch near Provo are almost lacking at Salt Lake City. This is also due to the recession of the original fault scarp and the obliteration of the clean-cut triangular spur ends. The Utah State Capitol stands on one of the spurs which projects unusually far beyond the present front of the Range. It seems extraordinary that this spur was not cut off by the main fault but field study shows that the fault does not cut across it. Field observation also shows that the front of the range has receded almost a mile. This has resulted in the development of a rock pediment,[55] now concealed under the alluvium, except where transected by streams.

The geological structure of the Wasatch is far too complicated to be described in a few words, or to be grasped in a hasty glance from an airplane.

Emigration Canyon, through which the air route runs, is cut in Jurassic limestones which are strongly folded and stand up almost vertically in the canyon walls. The crest of the southern Wasatch viewed from a high altitude over the Great Basin presents a fine panorama of matterhorn peaks[56] and snow-filled cirques. The valleys near at hand, however, show no indication whatever of having been glaciated, except in their uppermost parts where there is an occasional suggestion of an amphitheatre-like form.

The vegetation cover on the Wasatch Mountain front is very sparse and patchy, being concentrated largely in the hollows and ravines and on the northern and eastern slopes where the soil is less completely dried out by the sun.

Lake Bonneville Shorelines. Views to the north while passing over the Wasatch Mountain front show two remarkable terraces near the foot of the range (Fig. 59). These two terraces mark the old shorelines formed by Great Salt Lake which, during glacial time, stood at this level, some 900 feet above its present elevation. The climate was then much more humid than it is now. The large lake existing in Utah at that time has been given the name Lake Bonneville, after one of the early explorers in this region. Lake Bonneville discharged northward into Snake River and the highest shoreline, known as the Bonneville shoreline or terrace, represents that level. The outlet was rapidly cut down to a hard rock ledge which maintained the lake at a lower level for an extended period during which the Provo terrace was formed. This is called the Provo level because at Provo, Utah, a few miles south of Salt Lake City, there is a very distinct and wide terrace which represents a delta built into the lake at that time. Other broad alluvial plains, originally deltas built into Lake Bonneville, extend fanwise from the mouths of several canyons. These have now been deeply trenched by the present streams. Their lower slopes have been extensively irrigated and now constitute a zone of verdure between the barren mountain front and the even more barren desert to the west. This belt has been called the oasis of Utah. Here along the front of the Wasatch most of the population of Utah is to be found,

[55] Rock pediments, usually covered with a thin veneer of sand or gravel, fringe many of the basin ranges. This means that there has been a great deal of erosion since the range was first uplifted. A rock pediment is virtually a local peneplain at the base of the block mountain.

[56] A matterhorn peak is a sharp pyramidal peak (like Mt. Matterhorn in the Alps). It is formed by the working headward of several cirques into a mountain mass until its original round form is destroyed. This represents therefore an advanced stage of glaciation.

FIG. 60. The level floor of the Great Basin at Salt Lake City. Antelope Island, a block mountain standing in Great Salt Lake, in the distance.

and practically all of its cities, notably Ogden, Salt Lake City and Provo. In this rich area sugar beets and alfalfa are the outstanding crops but truck gardening and fruit growing are also important, as well as dairying.

The Salt Lake City airport occupies part of the bed of Lake Bonneville, two or three miles west of the city, beyond the Jordan River. From the flying field Antelope Island, one of the basin ranges surrounded by Great Salt Lake, is the most conspicuous feature to the northwest (Fig. 60). To the east the bold front of the Wasatch dominates the entire view (Fig. 61).

Great Salt Lake. West of Salt Lake City the air route for a few miles passes over rich cultivated fields watered by irrigation ditches and then for about 20 miles

FIG. 61. The western face of the Wasatch Range, seen from the Salt Lake City Airport. Alluvial fans extend from the front of the range into the Great Basin.

FIG. 62. The northern end of Oquirrh Range, seen from over Great Salt Lake. The view shows the higher or Bonneville Terrace and the lower or Provo bench of Lake Bonneville.

crosses over part of Great Salt Lake. At Salt Air, a resort on the shore of the lake, may be seen some of the evaporating basins designed for the recovery of salt. Rows of glistening white piles of salt attract the eye. The oscillation in level of Great Salt Lake is clearly shown by the many beaches and the broad plains of white salt which fringe the lake shore. The Salt Air pavilion was originally built far out into the lake to avoid the obnoxious character of the actual shoreline with its decaying plant and animal life, but the recession of the lake has brought the shoreline almost to the same position. Beneath the waters of the lake may be seen giant ripple marks and ridges formed by the waves. Sea gulls are very common around Great Salt Lake and some will almost certainly be seen flying far beneath the plane.

The Basin Ranges. The Oquirrh Range, lying to the south, now offers an exceptional opportunity to see both the Bonneville and Provo terraces from an altitude not much greater than their own (Figs. 62 and 63). Long sloping apron plains or alluvial deposits extend to the east. Near at hand is the Great Salt Lake Desert, a vast plain of glistening white, across which runs the highway and railroad.

FIG. 63. Oquirrh Range and part of Great Salt Lake Desert, a few miles west of Salt Lake City. On the side of the range may be seen the Bonneville as well as the Provo Terrace of former Lake Bonneville.

FIG. 64. One of the smaller maturely dissected basin ranges forming a promontory in Great Salt Lake, just west of Salt Lake City. Carrington Island in the distance.

To the north Antelope Island presents an interesting picture, rising abruptly out of the lake waters much as the other ranges do from the basin floors. Still farther away other blue ranges appear in the distance on every side. Here and there promontories project out into the lake like headlands into the sea (Fig. 64), producing many of the effects of a submerged coast.

FIG. 65. Great Salt Lake Desert, showing old shorelines on east side of Toand Range.

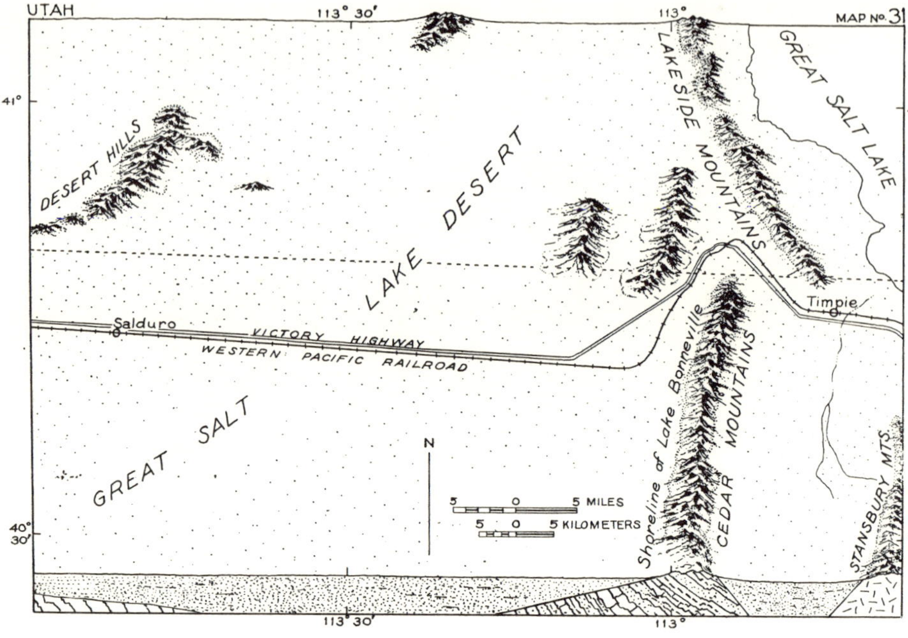

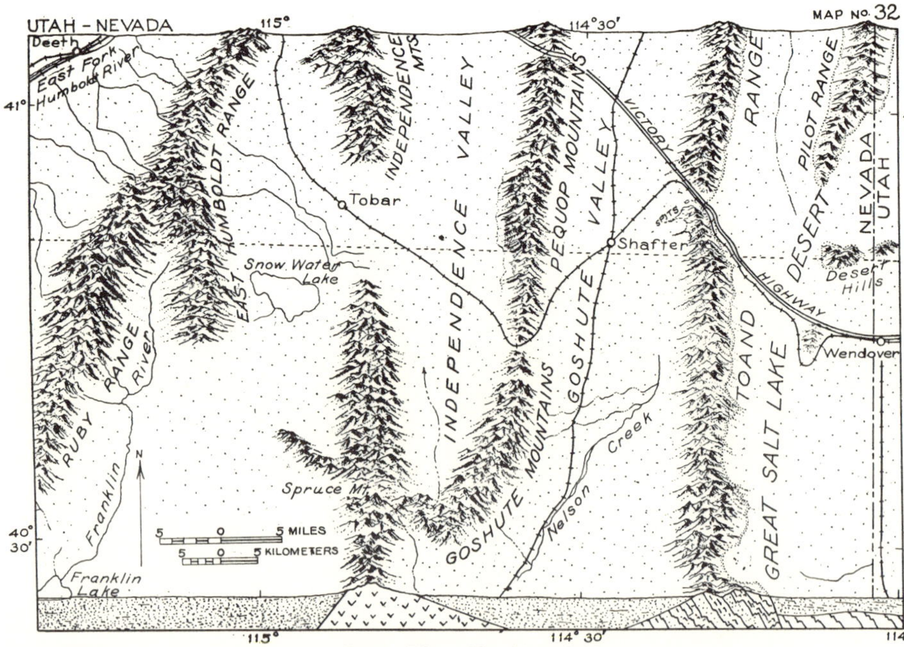

Route Maps No. 31 and 32

Route Maps No. 31 and 32

GREAT SALT LAKE DESERT TO RUBY MOUNTAINS

Great Salt Lake Desert. Beyond Great Salt Lake the Stansbury Mountains, lying south of the air route, display the Bonneville shoreline on their eastern and northern slopes. The route then crosses over the southern tip of the Lakeside Mountains. From their western base stretch long plains of alluvium. The Bonneville shoreline here is not very prominent. The Cedar Mountains lying to the south, with their growths of cedars, is an easily identified range. The easterly dip of the rocks in this range is clearly observed, and appears to be a continuation of the structure seen in the Lakeside Mountains to the north. The smaller ranges just northwest of Cedar Mountains present a beautiful display of terraces. Beyond these is the vast expanse of Great Salt Lake Desert (Fig. 65). Ripple marks and beaches of the former lake are everywhere present. Most of the surface is white salt. In some places there are sandy flats with a slight growth of vegetation, mainly clumps of greasewood, a salt-loving plant. Star-like spots of white devoid of vegetation form a regular pattern in the greasewood cover. These appear to be places where the salt has accumulated to a greater extent than elsewhere, perhaps being slight hollows in the surface of the ground.

Across the Great Salt Lake Desert, for 50 miles without a turn, run the Victory Highway and the Western Pacific Railroad. In all directions the mountain ranges rise abruptly above this plain, like islands in the sea.

The Basin Ranges. Immediately west of the Nevada-Utah boundary line (map No. 32), a number of ranges intercept the air route, necessitating steep climbing and often bumpy riding. Toand Range (Map 32) rises 3,000 to 4,000 feet above the desert plains. This range consists mainly of Paleozoic limestones, quartzites, and shales, with Tertiary lava beds and volcanic ash deposits on its western slope. The air route goes over its highest part, but the road and railroad find a pass a few miles to the north. Along the steep eastern base of this range are many remnants of the Bonneville shoreline. The Toand range, unlike the smaller basin ranges, supports considerable vegetation. On the summit are to be seen park-like meadows with scattered growths of trees.

The western side of Toand Range displays old beaches with interesting spits and other shoreline details. These are best seen just north of the air route. Goshute Valley is not like the desert lying to the east of Toand Range. It is brown in tone, but there are some green patches. Shafter, a railroad junction, lies in this valley and here may be seen a landing field (See Fig. 114), situated almost on top of one of the old shorelines.

The Pequop Mountains, with their easterly-dipping beds are next crossed. On the western side are shorelines which were formed by one of the smaller Quaternary lakes[57] of the Great Basin. White salt incrustations in the middle of Independence Valley are deposits left by this lake. Across this valley are several air route beacons which can be easily seen. Tobar, on the railroad, is readily identified. Snow Water Lake, like most playa[58] lakes in the Great Basin, varies greatly in extent and may at times be quite dry, being replaced by a salt flat. In this country there is almost no sign of human habitation, except the desert roads.

East Humboldt Range presents a splendid sight to the north and directly ahead lie the Ruby Mountains with their snowy peaks. These ranges all rise to elevations of almost 11,000 feet, which is 5,000 to 6,000 feet above the basin floor. They clearly display many glacial features on their summits, mainly cirques and trough-shaped

[57] Quaternary lakes are those which existed in glacial and post-glacial time, many of which have only recently disappeared.

[58] A playa lake is a temporary or intermittent lake, the term "playa" from the Spanish, meaning a shoreline, because of the numerous shorelines found around such lakes, due to their frequent change of level.

FIG. 66. The crest of Ruby Range, central Nevada, one of the highest of the basin ranges, showing features due to local glaciation.

valleys, and a sharp serrate skyline, altogether a more magnificent prospect than that part of the Sierra Nevada crossed by the air-line and shown on Map 37.

The Ruby Range is a typical Great Basin mountain range (Figs. 66, 70). It rises abruptly on all sides from flat valley plains or low, even slopes of rock detritus or "wash." The northern part of the range is granite of Cretaceous age. Flowing streams from the Ruby Range reach the Humboldt River in places. The lower mountain spurs are brown and barren, notably so on their southern and western sides. The northern slopes, being less dried out by the sun, are covered with vegetation.

In the valleys and on the alluvial slopes spreading westward from the Ruby Mountains are green flats and hayfields (Fig. 67). The streams dissecting the alluvial fans have developed wide trench-like flat-floored valleys. In these valleys may be seen meadows and cultivated tracts, with cattle and horses, all in great contrast with the desert aspect of the alluvial upland on either side.

FIG. 67. The Ruby Range, Nevada, seen from the west, showing snow-filled cirques and glaciated valleys. The cultivation of the alluvial fans in the foreground is made possible by irrigation.

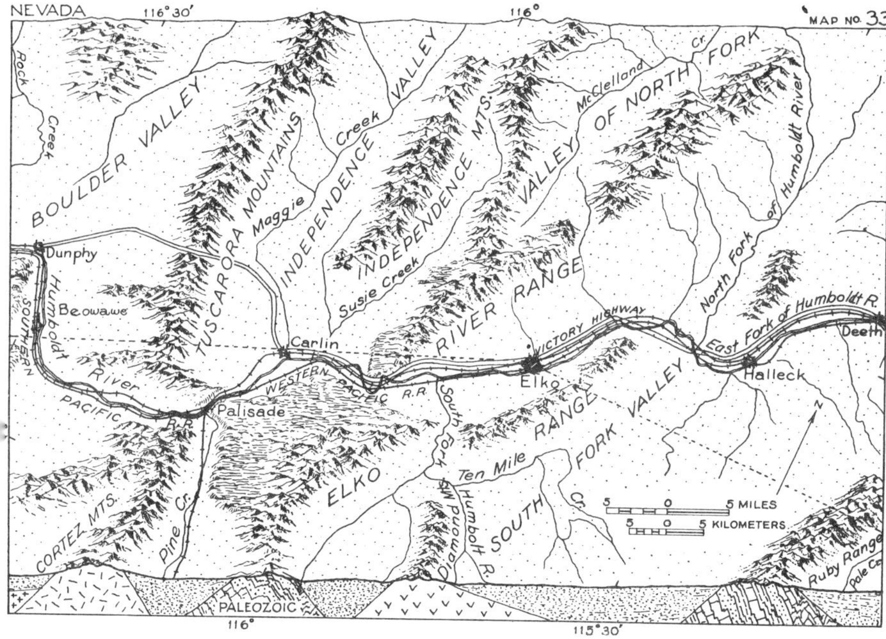

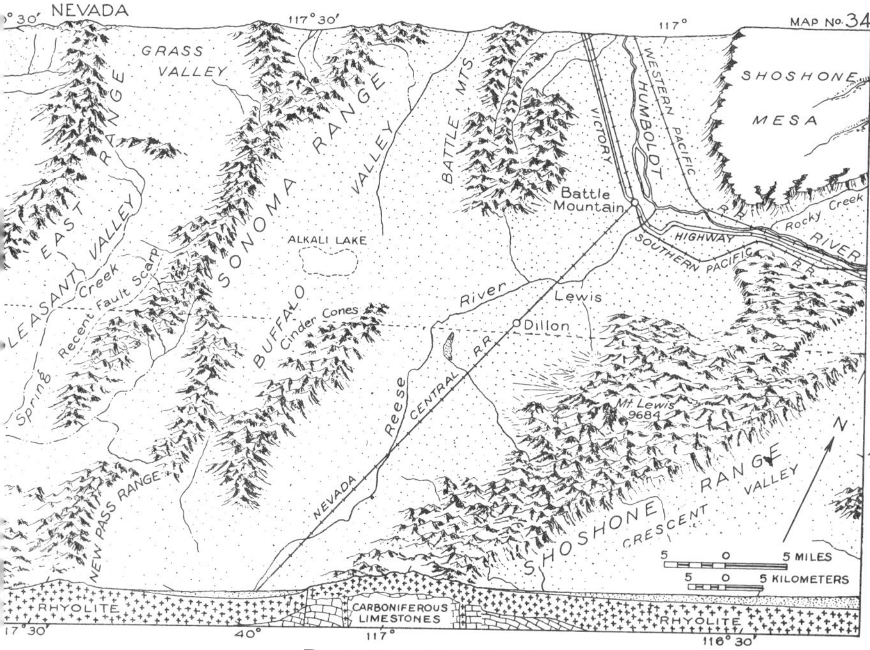

Route Maps No. 33 and 34

Route Maps No. 33 and 34

ELKO TO PLEASANT VALLEY

The Humboldt Valley. West of the Ruby Range (Maps 32 and 33) the rich valley of the Humboldt River comes into view. This broad, flat-floored valley with its ribbon of green is a true oasis in the desert. The Ruby Range, because of its great height and size, has an unusually heavy precipitation for this region and supports snow fields most of the year. The streams flowing out from this range, therefore, especially those on the western side, are amply supplied with water until the middle of July and the irrigation of the alluvial slopes at the foot of the range is therefore possible. Nearly one-fourth of this broad valley, between Ruby Range and Humboldt River is "self-irrigating" through seepage from higher irrigated lands. These self-irrigating lands are usually left in native grass which is naturally adapted to a scanty water supply. This provides pasture for cattle, or hay for winter use.

The Humboldt River flows on a flood plain flanked by low terraces or ridges, the remnants of a still higher flood plain. The Elko airport stands on such a terrace and other terraces rise above this level not far away. Wild grasses are grown for hay along the flood plain of the Humboldt River, and numerous haystacks are visible from the line of flight.

West of Elko the air route follows the valley of the Humboldt for 20 miles to Carlin, marked by a large white letter "C" on the hillside. To the south may be seen the Elko Range which consists mainly of Paleozoic rocks with Tertiary lava flows now forming a plateau. Through this plateau Humboldt River cuts its way at Palisade, so called because of the vertical jointing in the rocks, like the Palisades of the Hudson. This canyon with its steep walls of basaltic columns[59] is easily seen four or five miles south of the air route. South of Palisade a railroad runs southward up Pine Creek Valley 80 miles to Eureka, one of the most famous mining camps of Nevada, yielding large quantities of lead, silver, and gold.

Beyond Carlin the route crosses the southern end of Tuscarora Mountains, following one of the less important highways. A yellow radio beacon may easily be noticed on the way. Beowawe can be recognized by its rows of tall poplar trees. The fields along the river are abundantly irrigated and there are many haystacks. Just above the level of the main irrigation ditches the hill slopes are dry and barren. Off toward the south in Crescent Valley may be seen white salt flats. At some seasons it is likely that playa lakes exist there.

The Shoshone Range. The Shoshone Range (Map 34) over which the air route next passes after crossing the river at Beowawe, as well as Shoshone Mesa lying to the north, are plateaus of Tertiary basalts.

On the eastern side of Shoshone Range just beyond a salt flat may be seen vermilion and white hot spring deposits. These appear as a line or terrace and, like many others in Nevada, are situated at the base of the mountain front. Here, as elsewhere, the springs have probably risen along the line of the fault which separates the mountains from the valley.[60] Alluvial fans also extend from the eastern side of Shoshone

59 Columnar structure is common in lava flows, due to the jointing or cracking of the rock as it contracts during cooling. The individual columns are usually five or six sided and from a distance resemble rows of posts or piles.

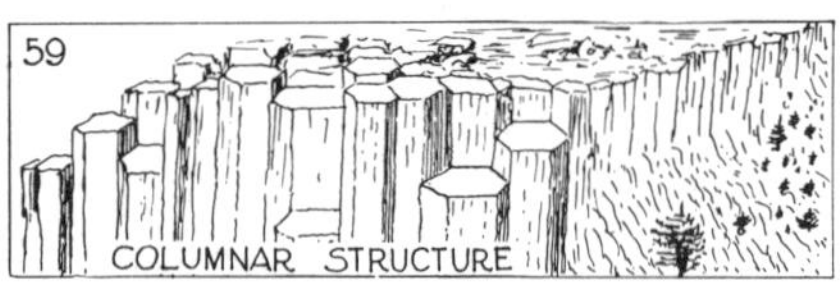

60 Many springs in the Great Basin are situated on fault lines. Hot spring deposits usually consist of white calcium carbonate, which is brought up in solution by heated waters passing through limestone along fault zones. Pink, red, and orange colors are due to the presence of iron in small quantities.

FIG. 68. Alluvial fans at the west base of Shoshone Range, view looking southward near Dillon, Nevada. This region is virtually devoid of all vegetation and the surface of the fans is intricately gullied by intermittent streams.

Range. The western side of the plateau is marked by a bold scarp, and the whole plateau apparently has an easterly dip.

Mt. Lewis, almost 10,000 feet high, lies a little to the south, and can be seen from the plane just before the railroad is crossed at Dillon. Long alluvial fans descend from the western slopes of Shoshone Range (Fig. 68), the apex of each one being at the mouth of a gulch. Several mining camps are visible in the valleys, with long streams of white refuse from the concentrators, covering the valley floor.

Young Cinder Cones. Beyond the Reese River Valley there is a small range along whose northern slopes may be seen a number of cinder cones (Fig. 69) adjacent to Buffalo Valley. These cones seem to be made up of red scoriaceous ash, and are apparently situated on the fault line at the base of the range. They are

FIG. 69. Cinder cones, on fracture zone, south side of Buffalo Valley, Nevada.

FIG. 70. The east side of Ruby Range, showing relatively straight baseline, and very abrupt descent to the plains.

moderately dissected, but they still retain their characteristic conical shape. In fact they are the best examples of volcanic forms to be seen along the air route.

These cones were visited by members of the King Survey[61] which explored this region between 1867 and 1873. They are described by Arnold Hague in the following words:

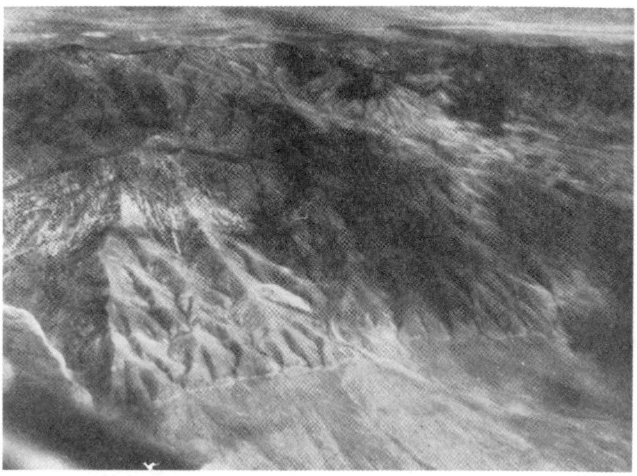

FIG. 71. The Sonoma Range, Nevada, seen from Pleasant Valley. The white streak at the base of the range is the recent fault scarp, dating from the earthquake of 1915.

[61] The King Survey, otherwise known as the 40th Parallel Survey, studied a belt of country across Colorado, Utah, Nevada and California with a view to determining the character of the mineral resources along the line of the Pacific Railroad, which had just been completed.

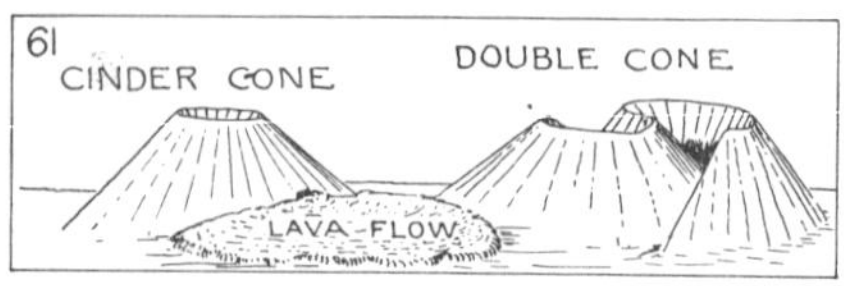

(*Photo by Professor Paul F. Kerr, of Columbia University, made in 1932.*)
FIG. 72. Sonoma Range, Nevada, seen from Pleasant Valley. The white line at the foot of the range marks the fault scarp produced by the earthquake of 1915.

"Skirting the western base of the table are a number of low rhyolitic hills and cones, scattered about in a most irregular manner, which are evidently of later origin than the great body of the rhyolite table. They appear to be local centers of eruption, crater-cones through which the last remnants of volcanic material here reached the surface. Two such isolated cones, rising nearly 300 feet in height, form prominent landmarks. Among these cones are a number of basaltic eruptions, 200 to 300 feet in width which are later than the rhyolites, breaking through them and forming hard overlying beds. Two especially prominent cones rise out of the volcanic debris

(*Photo by Professor Paul F. Kerr*)
FIG. 73. Base of the Sonoma Range, Nevada, showing the fault scarp with 16 foot displacement, formed in 1915.

about five or six miles south of the extreme northern end of the mountains. These cones present a very symmetrical outline, with a nearly circular base, and steep regular slopes. The lower two-thirds of the cones consist of light gray rhyolite, while the summit is composed of dark basaltic rock, which after the rhyolite had ceased to pour out, reached the surface through the same vents, building up the cone originally formed by the earlier flows. The contrast between the two varieties of rock is very distinctly marked, the basalt extending down the slope over the rhyolite in well-defined beds. From the mouth of one of these craters the last outburst has been a porous basaltic lava, which has not only poured down the sides of the cone, but has run out for a quarter of a mile on the plain.''

Recent Fault Scarps. On the west side of Sonoma Range which is next crossed, there is to be seen a recent fault scarp formed by an earthquake in 1915 (Figs. 71, 72 and 73). It appears as a very distinct white line, almost like a road, skirting the base of the range and running over the heads of the alluvial fans which descend westward into Pleasant Valley. Within a week, after the earthquake occurred, the locality was visited by Professor J. Claude Jones, of the University of Nevada, who described the rift and other effects of the earthquake in the Bulletin of the Seismological Society of America. To the geologist a feature of this kind is of unusual interest. It serves as almost direct proof that the range, along whose base it occurs, was originally formed by faulting. For most of the basin ranges such direct evidence is lacking. In most cases a straight line suggests the presence of a fault and this (with certain accompanying features, such as triangular facets) is almost all the available evidence for the fault origin of the basin ranges.

This part of the Sonoma Range is composed of rhyolite and basalt, the rocks dipping to the east. The western face of the range is an eroded fault scarp. The position of the fault can be followed along the western base of the range by its topographic effect where it crosses shoulders in the projecting ridges, and by lines of hydrothermal action.

The recent fault scarp formed at the time of the 1915 earthquake follows the old fault line very closely. There is, however, a tendency for the break to be somewhat west of the earlier fault plane, that is, further from the mountain front. This is due to the fact that where the fault crosses the unconsolidated deposits of the alluvial fans, a wedge of the alluvium clings to the mountain mass and is uplifted with it,[62] but where the rift crosses solid rock the old and the new fault traces are practically coincident. The recent fault appears at the surface as a fresh vertical scarp from five to fifteen feet in height, running for more than 22 miles along the junction of the bajada or alluvial apron and the base of the Sonoma Range.

62 The accompanying sketch shows diagrammatically the position of the recent scarp in relation to the mountain front which represents the earlier scarp. The mountain front, in places, has been worn back from the fault plane since the mountain was first uplifted.

62

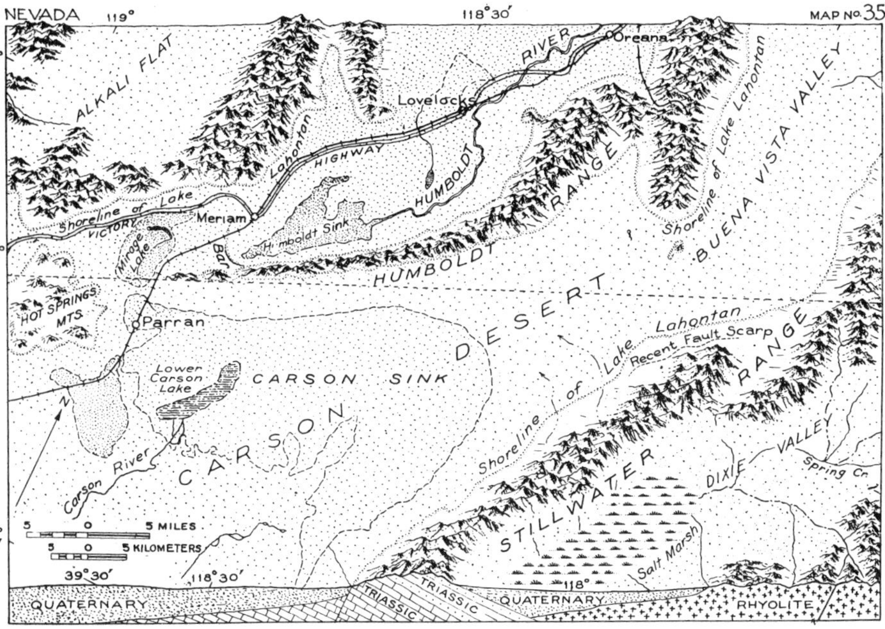

Route Map No. 35

THE CARSON SINK REGION

Former Lake Lahontan. The Stillwater or Pahute Range lies on the western side of Pleasant Valley (also called Dixie Valley) and forms the eastern rim of Carson Desert. Along its western base may be seen the highest shoreline of former Lake Lahontan. This lake, in many respects like Lake Bonneville of Utah, differed from it, however, in the important fact that it never had an outlet to the sea, or to any lower basin. It therefore oscillated much more in level and in consequence its highest shorelines are less emphatic in their appearance than those of the Bonneville region. Nevertheless, splendid bars and spits are to be seen, especially along the south base of Humboldt Range where the air route for many miles looks down upon the old beaches (Fig. 74). Alluvial fans built since the formation of the beaches occur at many places.[63] The beaches are being obliterated by being buried under alluvium and are also being worn away by streams, with the result that they appear intermittently. The most interesting of all the spits is the one which forms a great dam at the southern end of Humboldt Lake, sometimes called Humboldt Sink. This embankment of gravel from 50 to 125 feet in height above Humboldt Lake has been carried completely across the valley in such a manner as to suggest that it is an artificial structure intended to confine the drainage. At either end the main embankment

[63] Lake Lahontan and other Pleistocene lakes which occupied the intermontane basins of Nevada and Utah owed their size to an increase of precipitation as well as a decrease of evaporation during glacial time. In the glacial period no ice cap covered western United States, but the mountain glaciers expanded greatly, notably in the Wasatch and the Sierra Nevada, but also in the Ruby and East Humboldt Ranges of the Great Basin. It has been estimated that an

annual rainfall of 20 inches would restore Lake Lahontan. The above sketch shows the Lahontan beach dissected by recent gullies.

widens as it approaches the shore and forms heavy triangular masses. The best view of the entire bar is to be had at that point where the air route permits a view northward through a gap in the southern end of Humboldt Range. The main embankment declines in height from either end toward the center, and has been cut through at its lowest point by the overflow of Humboldt Lake. At its western end it is transected by the railroad.

On either side of the embankment there are extensive white salt flats which offer a strong contrast to the buff color of the sand and gravel of the bar itself. The accompanying figure (Fig. 75), based upon Russell's monograph on Lake Lahontan, shows in detail the appearance of these embankments.

Recent Fault Scarps. On the western side of Stillwater Range a fault scarp of recent geological age may be noted. From the air route it appears as a very straight line, going up and down over the features at the base of the range and cutting distinctly across the alluvial fans. It is by no means so young as the scarp on the east side of Pleasant Valley formed in 1915, for it was seen by I. C. Russell when he studied this region between 1880 and 1885. Several similar scarps of recent origin

FIG. 74. Lake Lahontan beaches at the base of Humboldt Range, Nevada. (U. S. Geological Survey).

are mapped in his Lake Lahontan report. Another one runs along the west side of Hot Springs Mountains (Map 35) and is responsible for the location of some of the springs. There is little difficulty in distinguishing between the scarps produced by recent displacement and the terraces formed along the old beach lines, although they have a superficial resemblance. The fault scarps go up and down over the features, whereas the beach lines go around them, keeping to the same level, like a contour.[64]

Carson Sink. Carson Sink is the name applied to the lowest part of the Carson Desert. During the winter and spring it receives a considerable supply of water from both the Humboldt and Carson Rivers and becomes a shallow playa lake 20 to 25 miles long. In unusually arid summers the water supply fails and the lake evaporates to dryness, leaving an efflorescence of salt several inches in thickness on the surface of the ground (Fig. 76). This disappearance of the water gave rise to the belief that the water sank into the ground, whence the name. The present small

64 The fault scarps seen from above are fairly straight but seen on a level with the eye they go up and down over the alluvial fans across which they cut. The beach ridges seen on a level with the eye are almost straight horizontal lines inasmuch as they represent an old water surface. From above, however, they appear quite irregular as they go in and out among the mountain spurs.

FIG. 75. Gravel Embankments at the west end of Humboldt Lake.

remnant of lake is usually called Lower or North Carson Lake. Its surface is now practically 500 feet below the level attained by Lake Lahontan at the time of its greatest extent. Many of the strand lines formed during its various stages encircle this lake, some at a distance of many miles.

The present annual precipitation of this region is only four inches per year, which renders this one of the dryest parts of the United States. When Humboldt Lake overflows and discharges by way of the lower Humboldt River through the bar it spreads out again to form Mirage Lake but this region is usually a blinding plain of white salt. Numerous channelways, however, are crossed by the air route.

Hot Springs. West of Carson Sink the Hot Springs Mountains or Kawsoh Mountains consist of flows and other volcanic rocks. Hot spring deposits are everywhere in evidence. The Lahontan shoreline fringes the base of the hills. On the eastern side of the mountain group the air traveller may note a radio beacon and a landing field situated directly on top of the uppermost Lahontan beach.

FIG. 76. The Stillwater Range, a view looking east over the salt flats of Carson Sink, Nevada, taken from the pilot house of a transcontinental plane. Beyond is Pleasant Valley and in the far distance is Sonoma Range.

To the west of this at Eagle Peak, in the Hot Springs Mountains, is a volcanic area of red buttes but no actual volcanic cones. Beyond on the western side of the range near the main highway are to be seen some evaporating basins used to recover the salt of this region. Views to the north from this point reveal the vast alkali flats (shown on Map No. 35) deposited on the floor of one of the smaller Quaternary lakes. This enclosed basin, because of its altitude, was never inundated by the waters of Lake Lahontan.

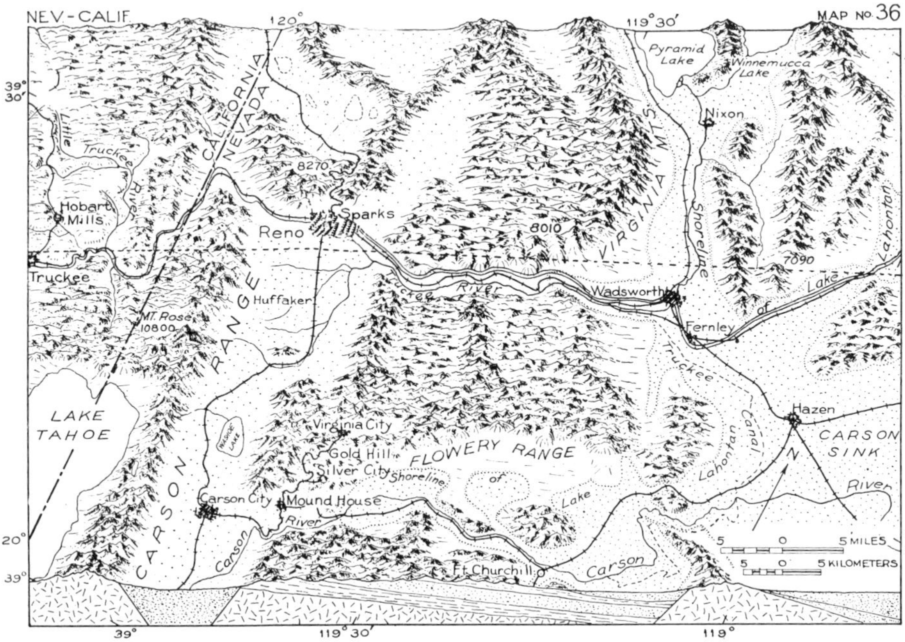

Route Map No. 36

THE RENO REGION

Lakes of Western Nevada. From Hot Springs Mountains (Map 35) westward the route roughly follows the main highway and railroad to Reno. Pyramid Lake, and Winnemucca Lake, sparkling blue bodies of water, appear far to the north, hemmed in by enclosing ranges. Both of these lakes occupy parts of the former Lake Lahontan basin. Pyramid Lake receives the water of Truckee River but has no outlet. It is therefore saline.

The Truckee Valley. The beautiful green tree-lined valley of the Truckee River comes into view at Wadsworth. This stream with its tributaries fed by the melting snows of the Sierra is responsible for the rich irrigated lands of western Nevada, notably the Truckee Meadows, near Reno. Here are to be found small irrigated farms where hay, grain, vegetables and various live-stock commodities are produced to supply the city markets.

Between Wadsworth and Reno the route follows the gorge of the Truckee River through the Virginia Mountains, where the river, the main irrigating canal, the railroad, and the highway all strive for a place. This canyon possibly follows a zone of faulting. Basalts and other igneous rocks make up the mountain masses on either side and along the canyon walls are many white hot-spring deposits.

Reno (Fig. 77) is a little city of about 19,000 population. Notwithstanding the prominence given its tourist and entertainment phases much of its more permanent prosperity rests upon the agriculture of the surrounding country. Situated in the Truckee Meadows, watered and traversed by the Truckee River, it nestles among a checkerboard of small farms, mostly about 40 acres in size, though tending to larger farms and stock ranches at a distance from the valley floor. Irrigation is the rule on the smaller farms, where the raising of alfalfa and hay constitutes the basis of the dairy industry. The farm homes are attractive, many of them constructed of stone

FIG. 77. Reno, Nevada, with the foothills of the Sierra Nevada in the distance. Many of the white lines crossing the picture are irrigation ditches.

and brick, set in pleasing plantings of trees, shrubs, and small family orchards, with much natural growth of cottonwoods and willows along the stream courses. This offers a lively contrast with the conditions prevailing in the arid basin-and-range section of Nevada. In most of the state away from the Humboldt and the Truckee Rivers the agriculture consists mainly of grazing lands with occasional narrow valleys that can be irrigated and where hay is grown to feed beef and sheep during the winter months. Near the larger towns of Elko (Map 33), Winnemucca (Map 36), and Lovelocks (Map 35), the type of farming is more diversified in order to supply the inhabitants with dairy and poultry products and vegetables.

All along the line through Nevada, sheep and beef cattle are out on the mountain ranges throughout the spring and summer months and are brought into the valleys in autumn for winter feeding. Away from the irrigated valleys the ranch houses are

FIG. 78. Hobart Mills, one of the largest lumber mills on the eastern slope of the Sierra Nevada, near Reno.

far apart and are always located near a supply of water. These ranch buildings are the headquarters for the men who take care of the cattle or sheep belonging to the ranch. The grazing lands usually extend into the mountains many miles from the ranch buildings. Because the vegetation is sparse, it is essential that the ranges be extensive in order to provide sufficient food for the grazing animals. Water is scarce and every spring and water hole is carefully conserved and becomes a center from which herds and flocks move to and from their pastures.

The Carson Range. The front of the Sierra Nevada at Reno is not a single well-defined scarp as it is in the Mono Lake region farther south, but consists of a number of partly detached blocks or fault splinters with intervening block basins.[65] The Carson Range is a block of this kind. Between it and the Virginia Range is the elongated block basin in which Carson City and Reno both lie. Between Carson Range and the main Sierra front is another block basin occupied by Lake Tahoe.

Westward from Reno the route passes just to the north of Mount Rose, the highest point of the Carson Range, with an altitude of almost 11,000 feet. Like the peaks of the High Sierra its summit has been glaciated. The cirques, separated by sharp-crested arêtes may readily be observed from the plane. The slopes of this range and the Sierra foothills are much more heavily forested than are the ranges of the Great Basin. One of the big lumber centers of the region is located at Hobart Mills (Fig. 78), easily seen to the north of the air route, which here follows the canyon of the Truckee River.

[65] The lower areas between tilted block mountains are known as block basins. A fault splinter is an irregular block which forms a small range at the foot of a larger block from which it has broken off.

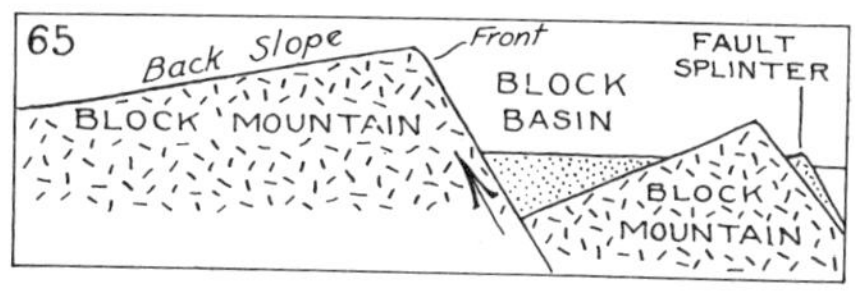

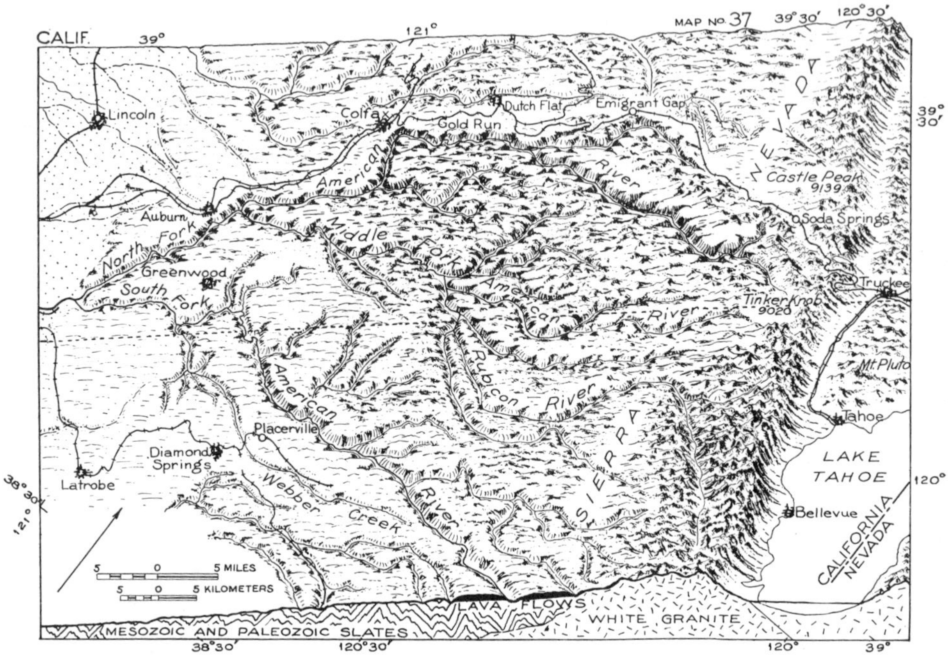

Route Map No. 37

THE SIERRA NEVADA

The Sierra Nevada Range. West of Truckee the main Sierra Nevada mountain front stands like a wall across the course of the air route. Above it the snowy peaks of the High Sierra rise to altitudes of 10,000 feet or more, some 3,000 to 4,000 feet above Lake Tahoe, distinctly visible to the south.

The actual elevation above sea-level is less than the crossing of the Laramie range, as the planes fly through a pass at 8,000 feet in good weather. When the pass is not free of clouds they rise to 12,000 feet to clear the summits. Landing fields in this area are not so conveniently established as in other ranges. There is no period, however, of more than ten minutes' flying when an emergency field is not within available range of the plane.

In bad weather and at night the airplanes follow close to the railroad route which is lighted by beacons, but in good weather the direct route takes them far to the south over a wild and unsettled part of the Sierra upland.

The Sierra Nevada is essentially a great tilted block, 125 miles wide, its upland surface sloping westward at about one-half degree or 50 feet per mile (Fig. 79). To the eye, however, it seems level. It may, for convenience, be divided into three north-south belts. The easternmost is the High Sierra. The second belt constitutes the true Sierra upland. The third, or westernmost belt, consists of a number of parallel ridges or foot-hills adjacent to the Great Valley of California.

The easternmost belt of higher peaks shows distinctly the effect of local glaciation. Although the cirques are not so large nor so clear-cut as those in the Uinta Mountains, nevertheless they serve admirably as examples of glacial action in summit areas. Many of them hold ponds or tarns. The air traveller looks down upon several of these immediately after the crest of the range is crossed (Fig. 80). Some of the lakes, such as Donner Lake, are undoubtedly held in by morainic accumulations

FIG. 79. The Sierra Nevada upland, west of Lake Tahoe. American River and its tributaries have cut deep canyons into the upland. The rocks in this view are mainly white granite.

but others are clearly rock basins. It is not always possible to judge just how far from the crest of the range the glaciers extended. It is obvious, however, from the broad appearance of some of the valleys, as for example the North Fork of the American River, that they were occupied by large glaciers in the same manner as were the Tuolumne and the Merced valleys farther south, in the Yosemite region.

The Sierra Upland. The top of the upland, with an average altitude of 6,000 feet, is open and park-like with pleasing meadows and forests extending one into the other. The various tributaries of the American River have cut deep canyons below the upland level which is preserved in many flat-topped remnants on the divides.

FIG. 80. The Sierra Nevada upland, west of Truckee, Nevada, near the headwaters of the American River. A few small glacial tarns and poorly developed cirques may be seen.

Close attention to the form and character of the valleys reveals bouldery ridges and other suggestions of moraines on the valley sides.

The Sierra Nevada block consists essentially of a basement complex made up in the eastern half of white granitic rocks, and in the western half of slates and other metamorphics of Mesozoic and Paleozoic age. Overlying these is a much later series of volcanic flows. On the top of the upland the light-colored granite outcrops extensively but farther west the outcrops of slate are less common because of the deeper soil. Much of this area is well forested although this is not the home of the big Sequoias. These occur 50 miles farther south. Extensive reddish or rusty patches reveal the effects of former forest fires and it is not unlikely that during the summer, one or more fires will then be raging somewhere on the upland within easy view of the air route.

The Western Sierra Foothills. In the foothill section, notably around Greenwood and Placerville, the mountains are maturely dissected and few flat-topped areas are in evidence. There is a strong contrast in the type of vegetation on the northeastern and southwestern slopes of the ridge-like divides. The slopes facing northeast, having the greater moisture because they are not dried out so much by the sun, support pine trees, whereas the slopes having a southern and western exposure are covered with sage-brush.

The towns in the foothill section of the Sierra are mostly small mining centers engaged in the production of gold through placer mining. The placer gold of the Sierra streams caused the great rush to California in 1849. This gold occurs in the gravels of the present day streams as well as in the so-called "higher gravels" of Tertiary age which lie under lava flows far up on the valley sides.[66] The mining operations, however, will hardly engage the attention of the air traveller until the Great California Valley is reached, at Folsom, to be described in the next section.

Agriculture in the Sierra. The first approach to the agriculture of California takes place when the farming country surrounding such towns as Colfax and Auburn comes to view toward the north. These typical mountain valleys lie on the western slope of the Sierra Nevada at elevations of 1,000 to 3,000 feet and are fully utilized for agricultural purposes. Soil and climatic conditions favor the growing of fruits, and the major farming activities center in the production of pears, peaches, plums, prunes, and other deciduous fruits. The farms are relatively small in size, averaging about 30 to 60 acres, well-cared for and home-like. Small fields of meadow and uncleared brushland vie with the orchards; all are irrigated, mostly from old ditches originally built to aid the early-day mining activities of the section. Thread-like streams of water can be noted running around the hills and across the little valleys. The towns, largely dependent upon the neighboring farms, though still to some extent catering to mining, are especially clean, well-built, and attractive. Paved roads radiate in every direction and serve to connect the towns and the farms. Especial note should be made of the clusters of packing sheds grouped at various points from which go to all parts of the United States the hundreds of tons of fruit produced in these valleys. Outside the irrigated area cleared fields provide grazing for sheep and cattle.

[66] During Tertiary time, lava flows from the higher parts of the Sierra poured down the valleys and buried the gold-bearing gravels. Later erosion formed new valleys, with the divides between them capped by the old lava flows.

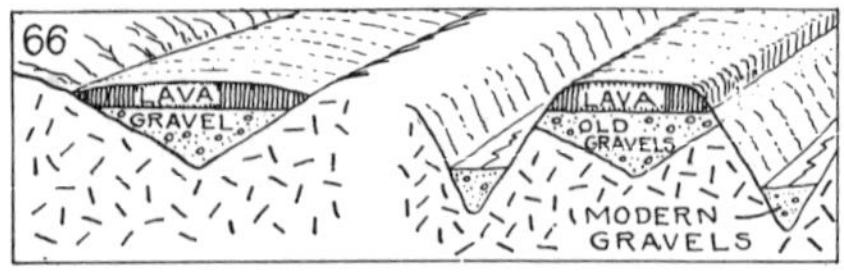

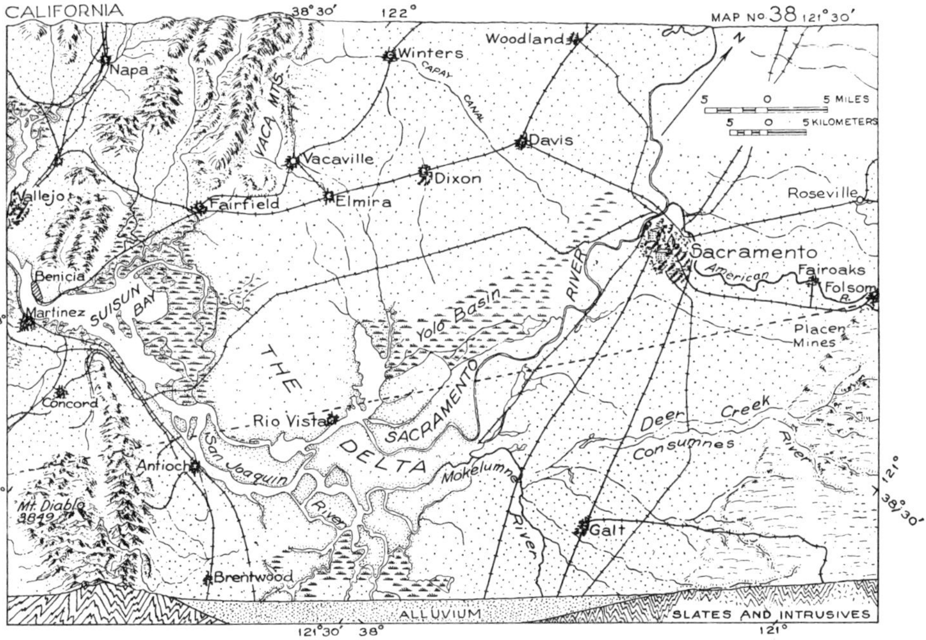

Route Map No. 38

THE SACRAMENTO VALLEY

Placer Mining. At Folsom the air traveller may behold some of the remarkable effects of placer mining on a large scale (Fig. 81). Extensive areas of the recent alluvial deposits spread by the Sierra streams upon the floor of the Great Valley have been reworked by dredges. The gravel, exhausted of its gold, has then been thrown up by the dredge in ridges, which, winding in a close scroll-like pattern across the country, mark the path taken by the dredge. In this manner the reworked areas have been changed from a land of agricultural productivity to a veritable desert. Between the ridges of gravel still stand stagnant pools of water, originally used to float the dredges. The observer, flying over this region, is almost certain to see a dredge at work. Close by these areas of waste land are rich green fields and orchards, clear testimony of the changes that have been wrought by this method of mining.

The Great California Valley. The Great Valley of California, lying between the Coast Ranges on the west and the Sierra Nevada on the east, forms a tectonic trough in which a series of sediments has accumulated, ranging in age from the earliest Cretaceous rocks to those of the present time.

In the center of the valley the Sacramento River pursues a winding course with numerous ox-bow bends and cut-offs. The stream has built up embankments one to 15 feet higher than the land on either side. The main channel is of very irregular depth and width and has not sufficient capacity to carry off the winter floods; in consequence, during high water much of the flow escapes through sloughs and crevasses into the lateral basins on the east and west, converting them for the time into extensive shallow lakes.

Far off toward the north, beyond the vast checkerboard which forms the floor of the Great Valley (Fig. 82), rises the group of volcanic peaks which constitutes the Marysville Buttes.

FIG. 81. Placer mining near Folsom, California, in the foot-hills of the Sierra Nevada. The view shows the dumps of gravel left by the dredges.

As one approaches the country surrounding the town of Fairoaks, situated about 10 miles east of Sacramento, and slightly to the north of the air route, a type of agriculture occurs which differs from that noted in the foothills of the Sierra. Though the topography is still rolling, the broader belts of agricultural land, the lower elevation (now about 675 feet above sea-level), and the change in soil types have combined to cause generous plantings of oranges, olives, and almonds. More grain fields, principally devoted to barley, are in evidence. Small poultry farms appear. But the country is still a country of homes, and of small-sized farms, and the landscape is sprinkled with neat, well-kept good looking sets of farm buildings, many supplied with electricity, the pole lines of the power company threading the country in all directions.

FIG. 82. The Great Valley of California near Sacramento, showing a vast checkerboard of fields. The scars of former stream courses may everywhere be seen meandering across the country.

FIG. 83. The Sacramento River near Walnut Grove in the Great California Valley. The Coast Ranges appear in the distance to the west.

The Sacramento Delta. The view from above Sacramento (Figs. 83 and 84), the capital of the state, is bound to focus upon the famous delta region, a broad, low-lying, level stretch of treeless country locally known as the "Netherlands of America." Mostly of peat soil, interspersed with sediment deposited by the waters of the Sacramento and San Joaquin Rivers, hot in summer and mild in winter, this broad body of land has been reclaimed from its original marshy state by levees and drains, into a series of islands; separated by water courses of various sizes. Here large farms are the rule, ranging in size from perhaps 160 acres to a section (640 acres) or more. The principal crops are potatoes, onions, sugar beets, barley, beans, asparagus, field corn, celery, and other truck crops, with the more suitable areas planted to substantial orchards of pears. The eye will be caught by the large tractors, plows, and other implements and machinery used in the cultivation of these lands, by the groups of

FIG. 84. The rolling topography in the Great California Valley, west of Rio Vista. This is a dissected alluvial surface, older than the present flat floor of the valley.

packing and canning plants built to care for the harvested products, and by the river steamers, large and small, which ply between the various islands, some of which can be reached only by some water route. The buildings tend to be rather flimsy in construction, often mounted on stilts as a safeguard against high water, and varying in adornment from elaborate landscaping to nothing at all, including lack of paint. There are exceptions, however, particularly where owners have erected homes for themselves and families. Large dwellings of substantial construction are in welcome contrast with the typical tenant and corporation-farmed holdings. Beyond the delta proper, that is just before the Coast Ranges are reached, the land rises to higher elevations and the soils change in nature, which in turn affects the character of the agriculture. Here extensive tracts planted to grain, beans, alfalfa, and other field crops are to be noted, with the accompanying utilization of live-stock, especially dairying and sheep. Large groups of buildings are indicative of big farms, and farms of several hundred acres each are not uncommon.

The Coast Ranges. For 50 miles the type of agriculture just described dominates the landscape. Surrounding the towns of Brentwood (Map 38), Concord, and Walnut Creek (Map 39), close to the foot of Mount Diablo (a notable high peak of volcanic origin outstanding against the hills of the Coast Range proper) there is a different type of agriculture. This region is characterized by numerous small family farms devoted to various deciduous fruits, of which pears, walnuts, and some grapes are noteworthy, interspersed with larger farms given over to the production of dry-farmed grain, and stock-raising on the hills. Alfalfa in limited amounts, poultry farms, and some dairying vie with the growing of field and fruit crops. Abundant acreages of tomatoes, raised for shipping to market and for the canneries, occur in this area. It is an attractive country of neat, well-kept homes. The climate is tempered by cool breezes off San Francisco and San Pablo Bays, which combine to form the broad body of ocean water lying to the north, above which rises Mount Tamalpais, and the low-lying hills upon which San Francisco is built.

From Walnut Creek (Map 39) westward the agriculture is not important. Leaving the valley proper the remainder of the trip is above the Berkeley Hills, covered with brush and trees, and though utilized for grazing and in the clearings for farming, the total is neither impressive nor important.

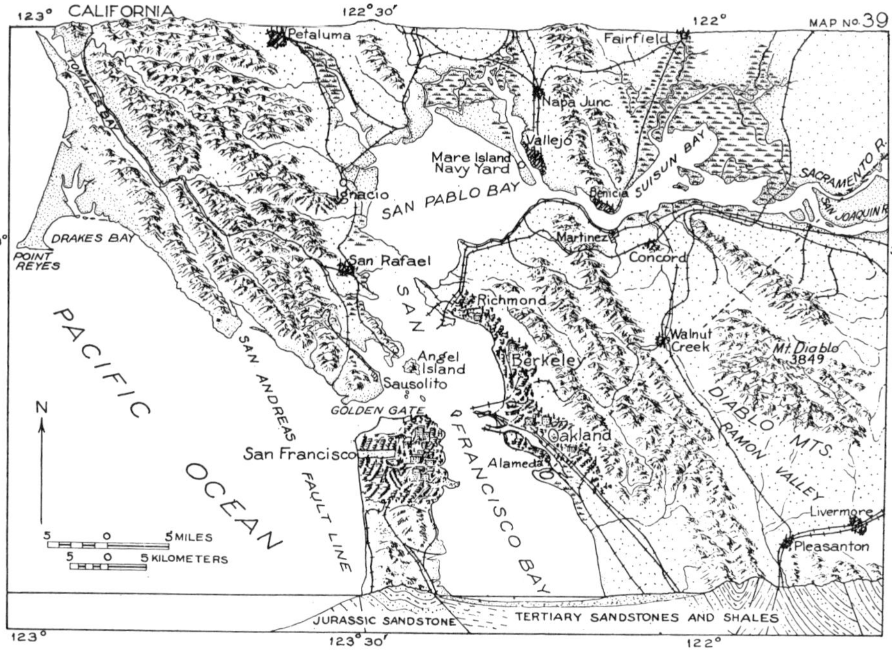

Route Map No. 39

THE SAN FRANCISCO REGION

The Coast Ranges. The region of the Coast Ranges about San Francisco consists essentially of three blocks, trending in a northwest-southeast direction. The easternmost block comprises a number of belts of hills lying to the east of San Francisco Bay. The middle block is a depressed area largely covered by the waters of San Francisco Bay. Its margins are occupied by many cities such as Oakland, Berkeley, and San Francisco itself. The westernmost block, next to the coast is, like the easternmost one, also made up of a number of parallel ranges. The first of these blocks is entirely crossed by the air route. The landing field at Oakland lies on the margin of the second block. The third block is to be seen only in the distance by the air traveller.

The Berkeley Hills. The Berkeley Hills block, as the eastern area may be termed, consists of several belts of hills rising to heights of almost 2,000 feet above the sea. The valleys between the ranges are fairly wide, notably Ramon Valley south of Walnut Creek, and they are deeply filled with alluvium so that they are flat-floored. The streams are now flowing in channels cut below the surface of these alluvial plains.

These valleys constitute important agricultural areas. The hills themselves, made up largely of Tertiary sandstones and shales, are maturely dissected and are little if any cultivated. The general aspect of the hillslopes is barren. The region is almost devoid of forest. Hill slopes and tops are bare of trees and the prevailing mantle of vegetation in uncultivated tracts is composed of the wild oat. Steep slopes facing the north and northeast away from the sun may be covered with a more or less dense growth of brush (Fig. 85). Several varieties of oak grow in such situations. Near the streams other trees, including willows, are found. To the air traveller the

FIG. 85. The Coast Ranges just east of Oakland and Berkeley, California. The northern and eastern slopes conserve the moisture and are clothed with vegetation. The sunny slopes are dry and barren.

dominant note in the landscape is complete dissection of the region with a prevailing tone of brown everywhere except in the gulches and on the shady slopes where the vegetation has a greener aspect.

The crest of the Berkeley Hills is a dividing line for the climate of this region. The climate of the area east of the crest is somewhat like that of the other interior valleys of the western United States; that of the west slope is like that of the coast. In the area east of the crest the summers are hotter and the winters are colder than in the area farther west. The sea breezes and fogs that temper the summer heat on the west slope have a greatly diminished influence on the east side of the Berkeley Hills. The annual rainfall at Berkeley and Oakland is about 27 inches and falls

FIG. 86. The city of Oakland, California, from the air. Mt. Tamalpais in the distance rises through the bank of fog which covers San Francisco Bay.

almost wholly in winter, the summer being rainless. Snow rarely falls, even on the highest ground.

San Francisco Bay. The Bay of San Francisco is a notable example of a great harbor formed by the influx of the sea into the low parts of a subsiding coast. Many other valleys of similar character between the coast ranges have not been inundated but otherwise are quite similar in structure and geological history. Most of these valleys are bounded or determined by faults, some of which are still active.

The most famous fault of the region is the San Andreas Fault just west of San Francisco. Horizontal movement along this fault plane in 1906 caused the great San Francisco earthquake. Throughout this region similar movements along the different fault planes occur each year, though few of them are of sufficient violence to cause destruction.

The Golden Gate Fogs. A phenomenon of particular interest to the air traveller in this region is the Pacific coast fog (Fig. 86). This fog, which extends for 50 miles or so offshore, moves in from the ocean with a westerly sea breeze, arriving on the coast about or soon after sunset and disappearing in the early forenoon. It penetrates inland not more than a few miles. It rarely exceeds 2,000 feet in height above the sea and consequently is more or less shut off from the land by the coast ranges except where the Golden Gate offers a free entrance. A traveller by plane, arriving late in the afternoon, is likely to see a great bank of cloud hanging over the Golden Gate and completely concealing the city of San Francisco at that time of day. These ocean fogs, with the onshore winds that accompany them, have been influential in causing the people of the region to seek the eastern side of San Francisco Bay for their residences in Oakland and Berkeley where there is less fog and less chilly wind.

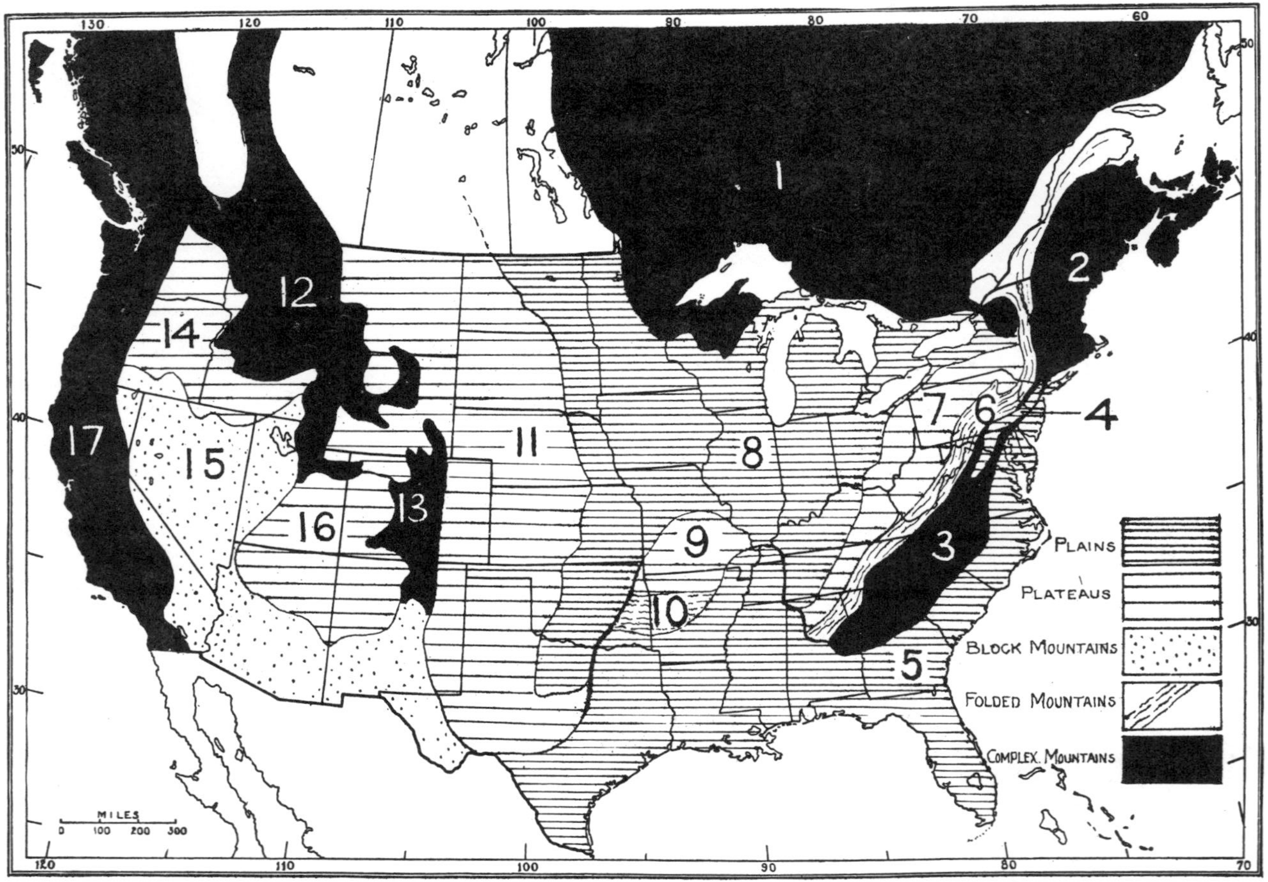
1
2
3
4
5
6
7
8
9
10
11
12
13
14
15
16
17
Plains
Plateaus
Block Mountains
Folded Mountains
Complex. Mountains
MILES
0 100 200 300

I. THE PHYSIOGRAPHIC PROVINCES OF THE UNITED STATES

General Statement. The air route from New York to San Francisco traverses 10 of the 17 major physiographic units of the United States. These are clearly shown on Figure 87 and also on Plate I. Besides these 10, an additional one, the Coastal Plain, is for a short time within view. Only six of the 17 major provinces are not seen and two of these, the Ozarks and the Ouachitas, are small and relatively unimportant. The reason most of the physiographic provinces of the United States can be crossed in a single traverse of the country is due to the simple and systematic arrangement of the several units and these in turn depend upon the geologic structure. Unlike Europe, which is broken into a checkerboard of many uplifted blocks and depressed basins, the larger structures of the United States are geologically intact. They trend in general north and south, parallel with the coast lines. In the belief that a concise summary of these major units will be helpful there follows a brief description of each of the provinces crossed by the air route, with especial emphasis upon the particular features and structures likely to come to the attention of the air traveller. Where deemed desirable an explanation of the geomorphology, that is to say the scenery, is introduced for the guidance of those to whom this subject is new, and it is assumed that the professional geologist will be tolerant of this and skip those portions which he will recognize as not intended for him. The smaller features, not to be identified on the generalized maps which accompany this chapter are left for consideration in the first part of this guide where the route is described in detail. It is suggested that this chapter as well as the other supplementary chapters be read at leisure before the journey itself is essayed. The detailed descriptions, on the other hand, are intended to be read while actually flying over the route.

The Atlantic Coastal Plain. Beginning, then, on the eastern seaboard, the first province encountered on this trip is the Atlantic Coastal Plain. This province is not crossed, but part of it is seen. South of New York City, in New Jersey, the Coastal Plain is narrow but it rapidly widens southward to 100 miles or more in the Carolinas. Long Island is a detached portion of the Coastal Plain. Still other isolated fragments occur on Cape Cod, and also make up the islands of Marthas Vineyard and Nantucket.

The Coastal Plain is a part of the sea floor which has been raised above water level. Its surface consists largely of loose sand, gravel, and clay, usually arranged in belts parallel with the coast. The weaker belts of clay have been worn away by streams, and the more resistant beds of gravel now form hills or ridges. Such ridges are termed cuestas (Fig. 88). The inner slope or inface of a cuesta is steeper than the back slope which conforms with the dip of the gravel layers. Occasionally the inface may be an abrupt scarp, as it is in the case of the Atlantic Highlands, south of New York Bay.

Where the Coastal Plain, after erosion, has been slightly depressed, then the lowlands previously formed become inundated and embayments result. Raritan Bay and Lower Bay of New York Harbor are examples of this. Long Island Sound is also a drowned part of the inner lowland, lying between the "oldland" of New England at the north and Long Island which constitutes a cuesta at the south.

FIG. 87. The Physiographic Provinces of the United States.

1. The Laurentian Upland; 2. The New England Province; 3. The Older Appalachians; 4. The Triassic Lowland; 5. The Coastal Plain; 6. The Folded Appalachians; 7. The Appalachian Plateau; 8. The Interior Lowlands; 9. The Ozark Plateau; 10. The Ouachita Mountains; 11. The Great Plains; 12. The Northern Rockies; 13. The Southern Rockies; 14. The Columbia Plateau; 15. The Great Basin; 16. The Colorado Plateau; 17. The Pacific Ranges. The air route traverses in order the provinces numbered 4, 2, 6, 7, 8, 11, 13, 12, 15, and 17. Part of province No. 5 comes into view.

Not all of the inner lowland has been drowned. That portion which forms the waist-line of New Jersey between Perth Amboy and Trenton is utilized by the great routes of transportation which run southward from New York. This belt, however, lies beyond the scope of our treatise.

The Triassic Lowland. The next unit is the smallest but nevertheless one of the most important physiographic provinces of the country, the Triassic Lowland. The name Triassic relates to the age of the rocks. These are largely red sandstones, containing, in the New Jersey area, dinosaur remains of Triassic time. This, however interesting as it may be, is of no great significance for our purpose. The important thing is that these rocks occupy an elongated basin, hemmed in on almost all sides by the much higher regions of complex crystalline rocks which in the New Jersey and New York area constitute parts of the New England Province. Farther south the Older Appalachians form the bordering highlands. (See Fig. 87.)

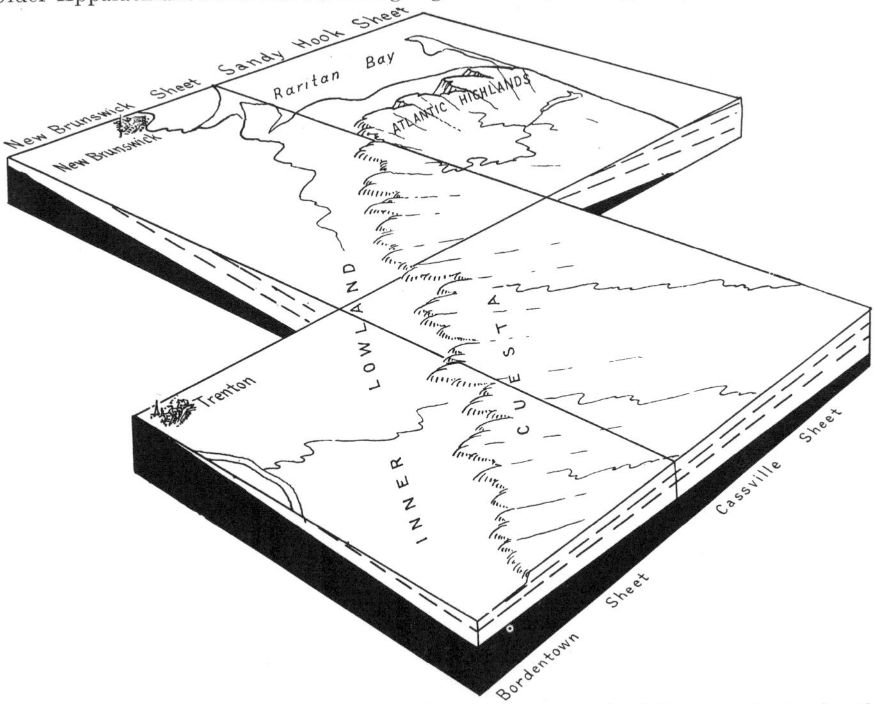

FIG. 88. The Coastal Plain of northern New Jersey, just south of the air route, showing the belt of hills forming the cuesta, and the inner lowland which traverses the state from Raritan Bay to Trenton.

The two southward-projecting extensions of New England which hem in the northern half of the Triassic Lowland on the east and west are known respectively as the Manhattan Prong and the Reading Prong of New England, the names being given from the cities situated at their southern tips. The projections of the Older Appalachians extending northward in a similar manner are known as the Trenton Prong and the Carlisle Prong, both of which are named after cities located at their northern ends. The Triassic Lowland is almost surrounded by these four prongs, but not quite. Owing to the fact that the Trenton Prong does not meet the Manhattan Prong, there is a gap between New York and Trenton along which the Triassic Lowland comes in contact with the Coastal Plain. And on the west there is a similar gap where the Triassic Lowland touches the belt of Folded Appalachians between

Reading and Harrisburg. On its western side the Triassic Lowland is cut off by a pronounced fault along which the Triassic beds have been dropped several thousand feet below the level of the crystalline upland. The Triassic beds therefore dip very strongly toward the west (Fig. 89).

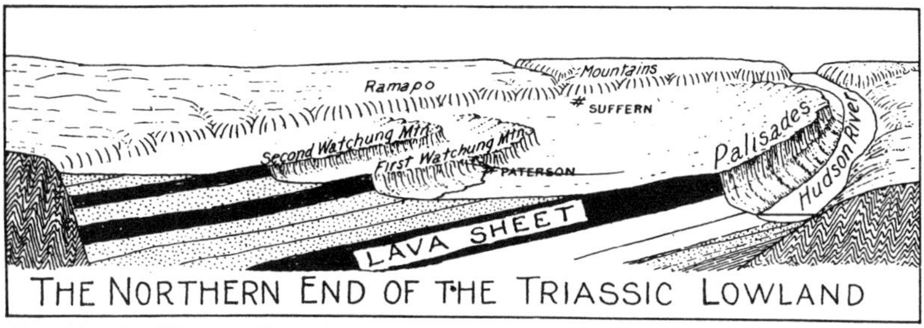

FIG. 89. The Northern End of the Triassic Lowland, west of New York City. In the distance is the Reading Prong of New England, known locally as the Ramapo Mountains. The geological section in the foreground shows the fault on the western side of the Triassic Lowland, and the strong westerly dip of the beds.

The Triassic Lowland is by no means everywhere a lowland. Its surface is interrupted by a series of trap ridges, due either to intruded sheets or lava flows interbedded with the sandstone. The sandstone, being more easily eroded than the igneous rocks, has been removed, leaving the more resistant trap sheets standing up as prominent features. The Palisades along the Hudson, the Watchung Ridges just west of Newark, and one or two lesser features of similar origin are all seen or crossed by the air route.

The New England Province. The third belt is the Reading Prong of the New England Province. Like the rest of New England it consists of complex crystalline rocks such as gneiss, schist, and granite, with some bands of limestone and shale which have been folded or faulted into it (Fig. 90).

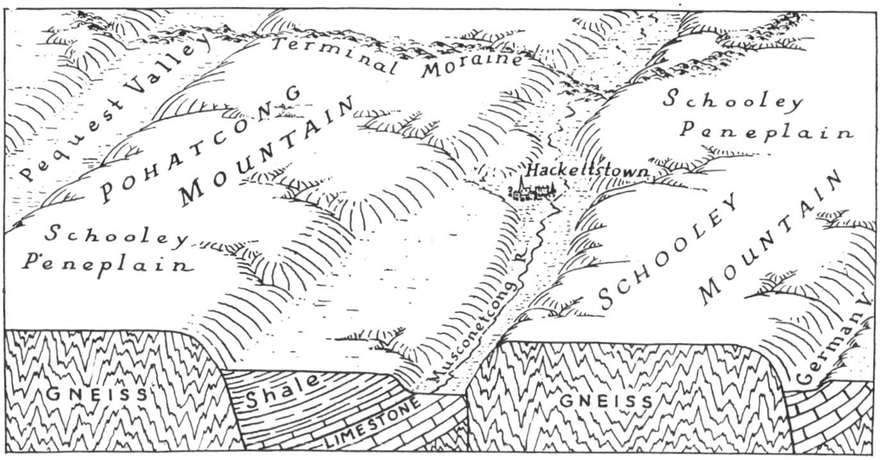

FIG. 90. Part of the Reading Prong of New England in western New Jersey, crossed by the air route. The geological section in the foreground shows that the higher parts of the region are underlain by gneiss and other crystalline rocks, and that the valleys have been eroded along belts of shale and limestone.

The New Jersey part of this belt is usually known as the New Jersey Highlands. Although it has a fairly rugged appearance it is not characterized by an irregular grouping of peaks but is in the nature of an upland. Its surface is a peneplain, a land area which at one time was worn down to sea-level and later raised to its present position. This elevation afforded opportunity to streams to remove some of the infolded weaker sedimentary rocks and these eroded belts now constitute the valleys of the region, such as German Valley, Musconetcong Valley, and Pequest Valley. Musconetcong Mountain, Schooley Mountain, and Pohatcong Mountain are terms applied to parts of the upland. The features of the Reading Prong, therefore, have a pronounced grain or trend, running in a northeast-southwest direction. The Delaware River crosses this crystalline belt at Easton.

The Newer or Folded Appalachians. West of the crystalline highlands, represented by the Older Appalachians and the New England Province together, is a belt of very much younger rocks which have been strongly folded. This constitutes the Newer or Folded Appalachian Province. The rocks making up this belt consist largely of limestones, shales, sandstones and conglomerates. Just as in the case of the crystalline rocks of the New England Province, so this folded belt suffered profound erosion until it was worn down to a plane with little relief. This plane beveled the different folds and also the different rock layers. It was then bodily raised and streams eroded the belts of weaker material, leaving the more resistant layers projecting as flat-topped ridges (Fig. 91).

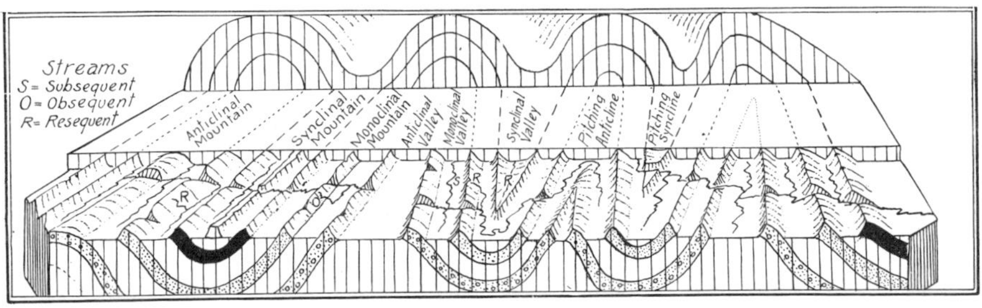

FIG. 91. Diagram showing the development of ridges in central Pennsylvania. The back block depicts the original folds; the middle block represents the region after peneplanation; the front block illustrates present conditions.

The ridges usually are parallel with each other but it will be noticed in Fig. 91 that where the folds pitch the resulting ridges zigzag back and forth. This zigzag plan of ridges in folded structures is better developed in Pennsylvania than anywhere else in the world. To the air traveller this is of especial interest because the air route passes over some of the most remarkable features of this kind.

The Folded Appalachian belt in Pennsylvania consists essentially of two parts, roughly separated by the Susquehanna River (Fig. 92). The eastern part is a complex basin area forking out toward the west in several prongs, the ends of which are crossed by the Susquehanna River above Harrisburg. Each prong represents the axis of a syncline pitching toward the northeast into the main center of the basin near Pottsville. On the eastern side a similar prong comes to an end at Mauch Chunk, the syncline in this case pitching toward the southwest, also toward the center of the Pottsville Basin. The folded belt west of the Susquehanna River consists mainly of a large anticlinorium which forks out toward the northeast and toward the southwest. At the north a series of prongs pitching toward the northeast may be noted on Figure 93. The axis of the most western anticline has been eroded to form Nittany Valley in which Bellefonte lies.

The easternmost of the ridges of the Folded Appalachians is Kittatinny or Blue Mountain, formed of a massive conglomerate which dips west. This ridge is cut across by several rivers in prominent watergaps, notably those of the Delaware,

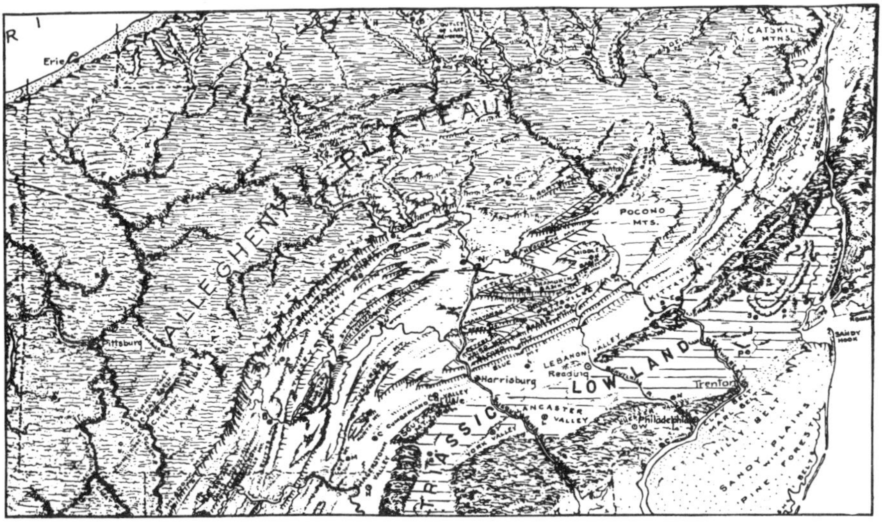

FIG. 92. Diagram showing the chief topographic features of New Jersey and Pennsylvania. The air route crosses the northern part of the Triassic Lowland, the Reading Prong of New England, the Folded Appalachians, and the Allegheny Plateau.

Lehigh, Schuylkill, and the Susquehanna. Two of these, the Delaware and the Lehigh, can readily be observed by the air traveller. East of Kittatinny Mountain is the Great Valley some 10 to 12 miles in width, underlain by limestones and shales. This valley lies just west of the belt of crystalline rocks, those of New England in the north and of the Older Appalachians in the south.

The Appalachian Plateau. The Appalachian Plâteau belt is a broad and very shallow syncline. Its width from central Pennsylvania to central Ohio is over 200 miles. On the eastern side the formations dip slightly to the west; on the western side slightly toward the east. The eastern side is marked by a bold escarpment

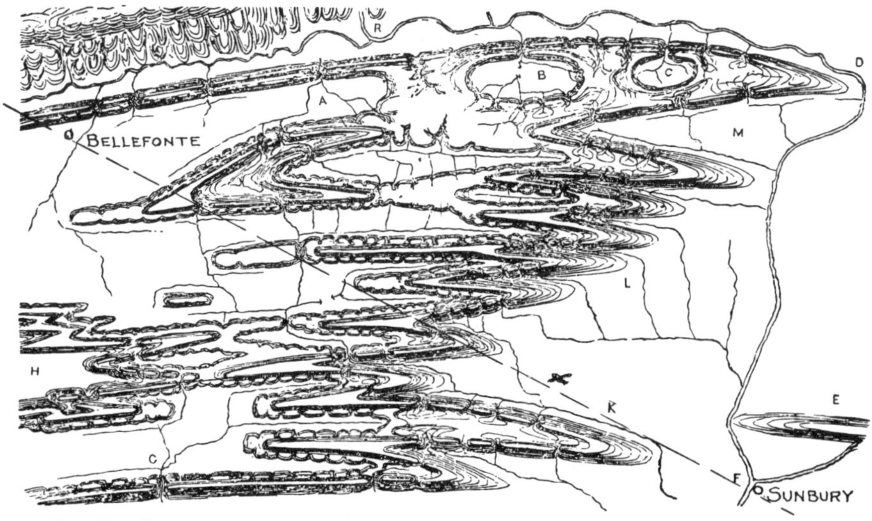

FIG. 93. Sketch map showing the series of plunging anticlines between Sunbury and Bellefonte.

rising almost 2,000 feet above the valleys of the folded belt. This is the Allegheny Front (Fig. 92). On the west, a much less pronounced escarpment crosses the state of Ohio from Cleveland southward. The northern part of the plateau, largely in New York State, has been glaciated, with resulting lakes, marshes, and drainage changes. The southern portion, represented in Pennsylvania and southward, is maturely dissected and constitutes one of the most rugged and least developed parts of the eastern United States. Practically all of the development that has taken place has been due to the occurrence of coal, oil, and gas within its borders.

The Interior Lowlands. The Interior Lowland belt, lying between the Appalachian Plateau and the Great Plains is not everywhere sharply separated from its neighbors. Nevertheless it has certain characteristics which distinguish it from the others. It lies at a lower altitude and has less relief than the provinces bordering it. In the area under consideration the effects of glaciation are paramount (Fig. 94).

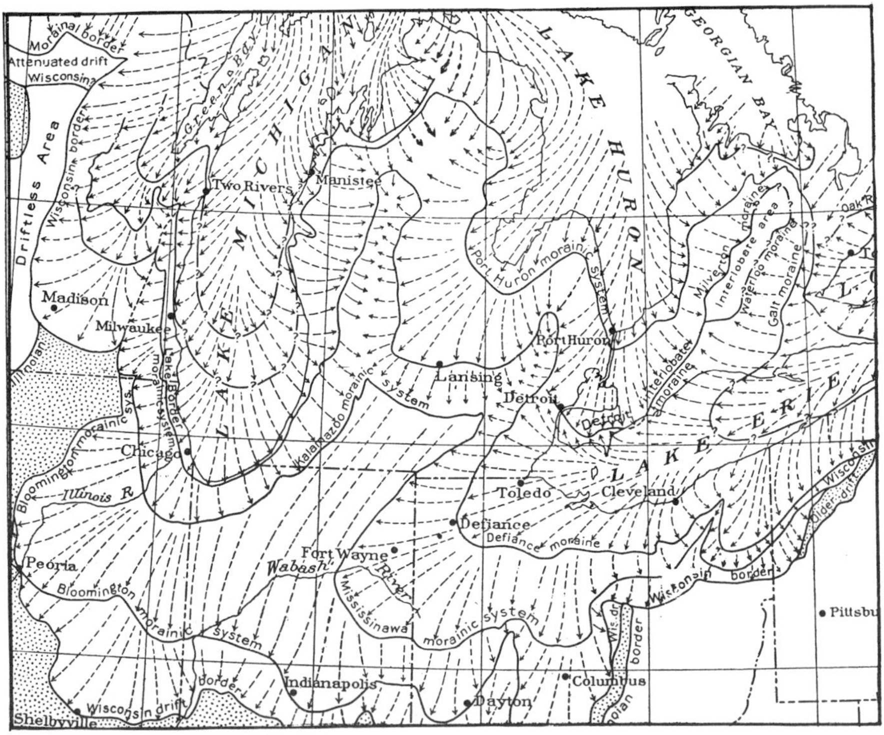

FIG. 94. Map showing the moraines deposited by the Lake Erie and the Lake Michigan lobes of the continental ice sheet. (U. S. Geol. Survey.)

Two lobes of the Continental ice sheet encroached upon the area crossed by the air route. One lobe advanced westward through the Lake Erie basin, the other southward through Lake Michigan, almost to the present position of the Ohio River. As the ice melted back successive moraines were deposited along the border of the ice sheet. These morainal belts now constitute the chief features in the topography of northern Ohio, Indiana, and Illinois. They are characterized by numerous lakes and marshes. Between the moraines are very level till plains or outwash plains of sand, devoid of lakes.

The dissected till plains of Iowa and eastern Nebraska are more rolling and possessed of a higher degree of relief than the till plains farther east. They represent earlier glacial deposits now deeply eroded. Practically all of the earlier lakes of these western states have disappeared.

The Great Plains. The Great Plains Province comprises, roughly, the area lying between the Missouri River and the Rocky Mountains (Fig. 95). It is a zone

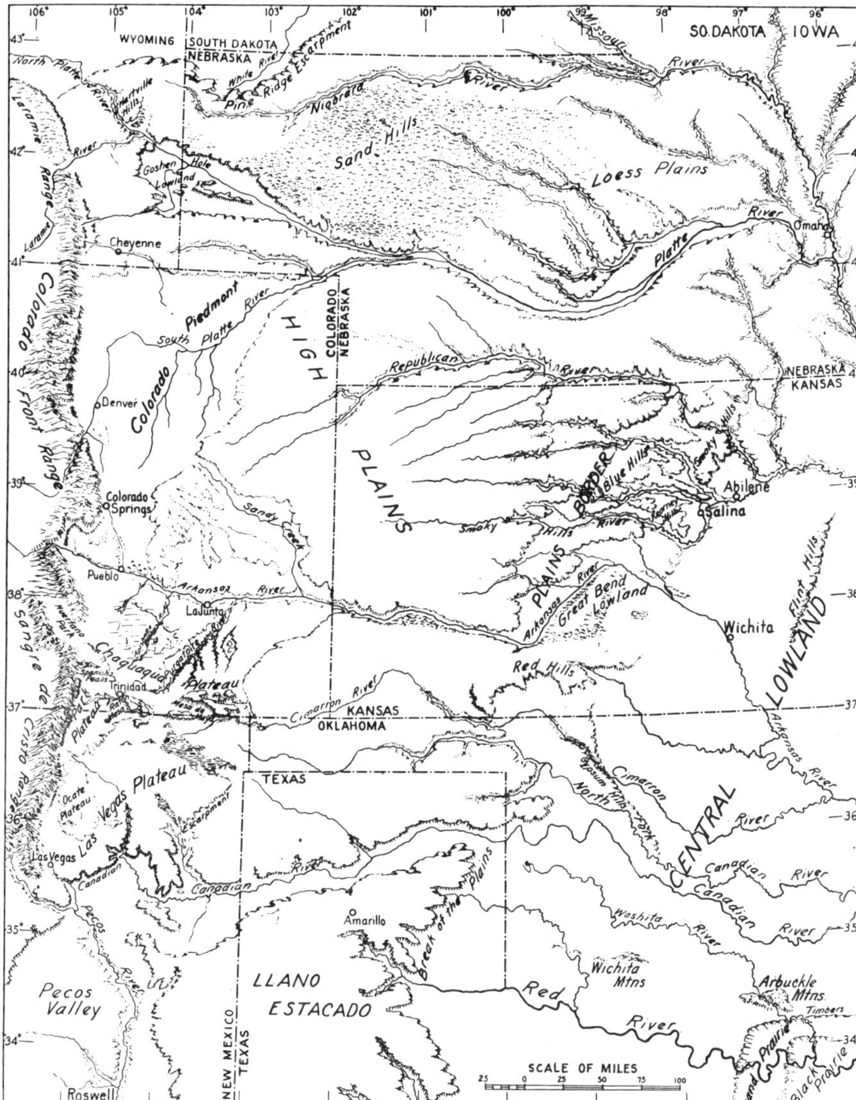

FIG. 95. Diagrammatic map of the Central Great Plains. (Drawn by Guy-Harold Smith for Fenneman: Physiography of the Western United States, McGraw-Hill Book Co., Inc.)

almost 500 miles in width which rises gradually from an elevation of about 2,000 feet at Omaha to over 6,000 feet at Cheyenne and extends the entire width of the United States from Canada to Mexico. In its eastern and more humid part, where it is drained and dissected by the numerous tributaries of the Missouri River, and in central

Nebraska and Kansas, the topography is much diversified and the landscapes are rolling. This is the Plains Border district, a type much more characteristic of Kansas than of Nebraska. In the Plains Border country the streams are cutting into Paleozoic and Mesozoic beds. The upland areas between the streams are largely covered with wind-blown loess which thickens westward and in central Nebraska, in the High Plains region, reaches a depth of 100 feet.

The High Plains constitute, at least in Nebraska, the largest and most typical part of the Great Plains. The surface of the High Plains is mainly the remnant of a great fluviatile plain or alluvial slope which spreads from the Rocky Mountains on the west to the Missouri River (Fig. 96). It is the eastern margin of this slope which has been dissected to form the Plains Border. Very few streams cross this large tract of country and because of the semi-arid conditions which prevail there, as well as because of the porous nature of the material making up its surface, it is largely undissected, and vast areas appear to be perfectly flat. In Kansas where the Ogalalla (Pliocene) formation makes up the surface, this undissected condition prevails and the surface of the plain represents the original surface of deposition. No hills or mesas rise above the level of the plains. In northwestern Nebraska and in southeastern Wyoming, however, the Arikaree gravels of Miocene age make up the region. These suffered dissection while the Ogalalla beds were being laid down farther south. This region, therefore, is not quite so flat and occasional mesas may be seen rising above the plain level. One of these stands north of the Cheyenne Airport.

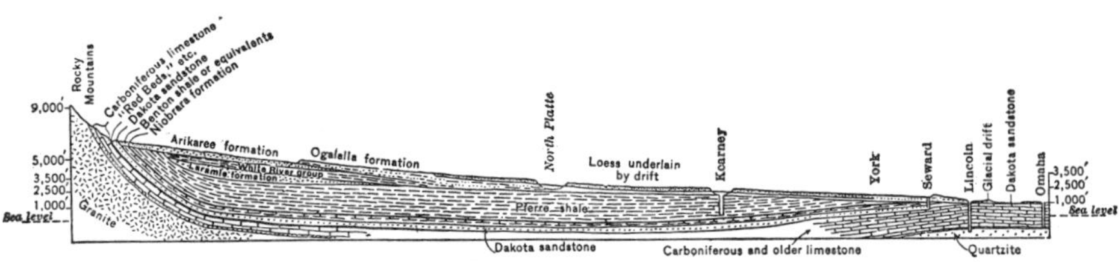

FIG. 96. Geological section through the Central Great Plains along the line of the air route. (Darton, U. S. Geological Survey.)

Much of the High Plains surface is covered with wind-blown material, both sand dunes and loess. North of the Platte River in Nebraska there is an area of 24,000 square miles where dune topography predominates. Some of these sand hills are several hundred feet high. Between the dunes open meadow lands occupy a small part of the area. There are also many lakes and ponds occupying depressions between the dunes. Over broad tracts the dunes are still drifting; elsewhere they are more or less fixed, depending upon the rainfall, which varies from season to season. In rainy summers most of the area is well grassed and supports a grazing industry. A large part of this tract of country is visible north of the air route in central Nebraska. East of the sand hill district the interstream areas are flat because of the deep covering of loess which conceals the older and more uneven surface.

The Platte River, formed by the union of the North and the South Platte, traverses the High Plains in a valley which lies slightly below the plain level. The North Platte, with its well-defined channel filled with a fairly large volume of water, and the braided South Platte, made up of a complex network of interlacing channels, constitute features of especial interest to the air traveller while crossing the western part of the High Plains.

The western margin of the High Plains at the base of the Rockies has in some places been eroded to form a lowland or piedmont area. The Colorado Piedmont, in which Denver lies, is the largest and best developed example. An irregular escarpment descends from the High Plains to the lowland floor. The Goshen Hole in western Nebraska is a basin of this type which, however, does not reach westward to the mountain front. At Cheyenne it is worth while to note that no lowland has been formed. The High Plains there rise gradually to the mountains and overlap their flanks, forming what is sometimes called a "land bridge." This bridge serves as the route over which the Union Pacific Railroad and the Lincoln Highway cross the northern end of the Southern Rockies. To the north of the land bridge and to the south of it there is a lowland between the mountain front and the scarp which rises to the High Plains.

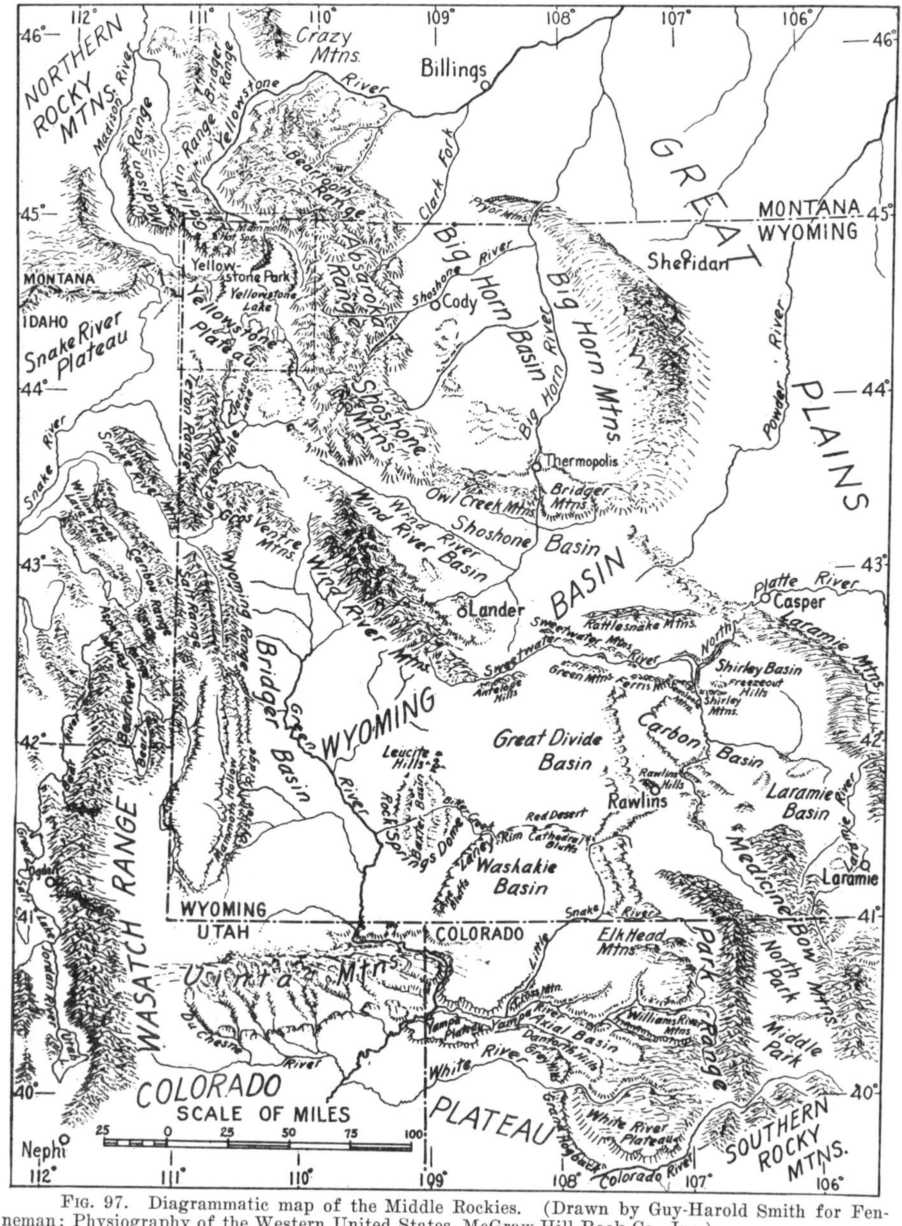

FIG. 97. Diagrammatic map of the Middle Rockies. (Drawn by Guy-Harold Smith for Fenneman: Physiography of the Western United States, McGraw-Hill Book Co., Inc.)

The Wyoming Basin (Fig. 97) lying between the Northern and Southern Rocky Mountains may be considered an extension of the Great Plains. There is a wide gap between the Bighorn Mountains and the Laramie Mountains which connects the two. The Wyoming Basin is essentially a broad basin in structure, but many uplifts, both around its margin and in its center, give it great diversity. It actually consists of a number of basins. It should be noted that some of the basins such as the Laramie Basin, Carbon Basin, Bridger Basin and Washakie Basin, are structural synclines, but certain other basins, like the Baxter Basin, are low anticlines in structure, breached at the summit and eroded to give a basin topography. Structural domes of this type are numerous in Wyoming. Some constitute great mountain ranges, like the Wind River Range, the Bighorn Mountains, and the Laramie Mountains, whereas many others are hardly to be detected in the topography. These smaller ones, like the Teapot Dome, for example, north of Rawlins, are the loci of important oil pools.

The presence of so many large domes, most of them maturely dissected, accounts for the many hogbacks and escarpments which dominate the scenery of the region.

The Southern Rocky Mountains. The Southern Rocky Mountain Province (Fig. 98) consists of a complex of north-south trending ranges concentrated mostly in Colorado but extending northward into Wyoming and southward into New Mexico. The northward projecting ranges are the Laramie Range on the east, the Medicine Bow next to the west, and the Park, the most westerly of all. Between the Medicine Bow and the Laramie Range is the Laramie Basin, a broad syncline underlain by Paleozoic and Mesozoic beds. Around the flanks of these three ranges the sedimentary formations have been turned up to form hogbacks which vary much from place to place, appearing as sharp narrow ridges or, where not so steeply dipping, as crenulated escarpments. In some localities they are entirely missing. The summit of the Laramie Range is a rolling upland, the granite core of the range having been truncated by a peneplain. The Medicine Bow Range also shows an even summit level, above which rise higher serrate snow-capped peaks. Partly detached from the Medicine Bow Range is Elk Mountain which forms its northern extremity. The many details of the Colorado Rockies must necessarily be omitted from this description.

The Northern Rocky Mountains. The mountains here included in the Northern Rocky Mountains are sometimes distinguished separately as the middle Rocky Mountain Province (Fig. 97). The ranges bear little definite relation to each other and as a whole the province can not be considered a well-defined unit. The Bighorn and the Wind River Mountains are both structural uplifts, eroded at the summit and flanked by the upturned edges of the bordering sedimentary formations. The Wind River Mountains can easily be seen from the air route across southern Wyoming but the Bighorns are too far away. Of especial interest to the air traveller is the Wasatch Range because the air route goes directly over its crest at its highest part. The range itself trends north and south and, like the Tetons in the Yellowstone Park area and the many smaller ranges lying between, it is a block bounded by a fault on one side. For that reason the Wasatch Range is sometimes included in the Great Basin Province with other basin ranges. The western face of the Wasatch, at whose foot the fault runs, rises steeply from the floor of the Great Basin like a wall 5,000 to 6,000 feet and more in height. Numerous canyons dissect the front of the range. The back slope is also deeply dissected into a rugged mountainous area with no suggestion of an inclined plane or even of a former rolling surface, except at rare intervals.

The Uinta Mountain uplift is a broad and high dome trending east and west. It is therefore rather anomalous among so many uplifts which run north and south. It is approximately 150 miles long and 30 to 40 miles wide and has many peaks more than 12,000 feet high. A tremendous thickness of sediments has been worn from the crest of the arch but nevertheless sedimentary beds still cap the summit which

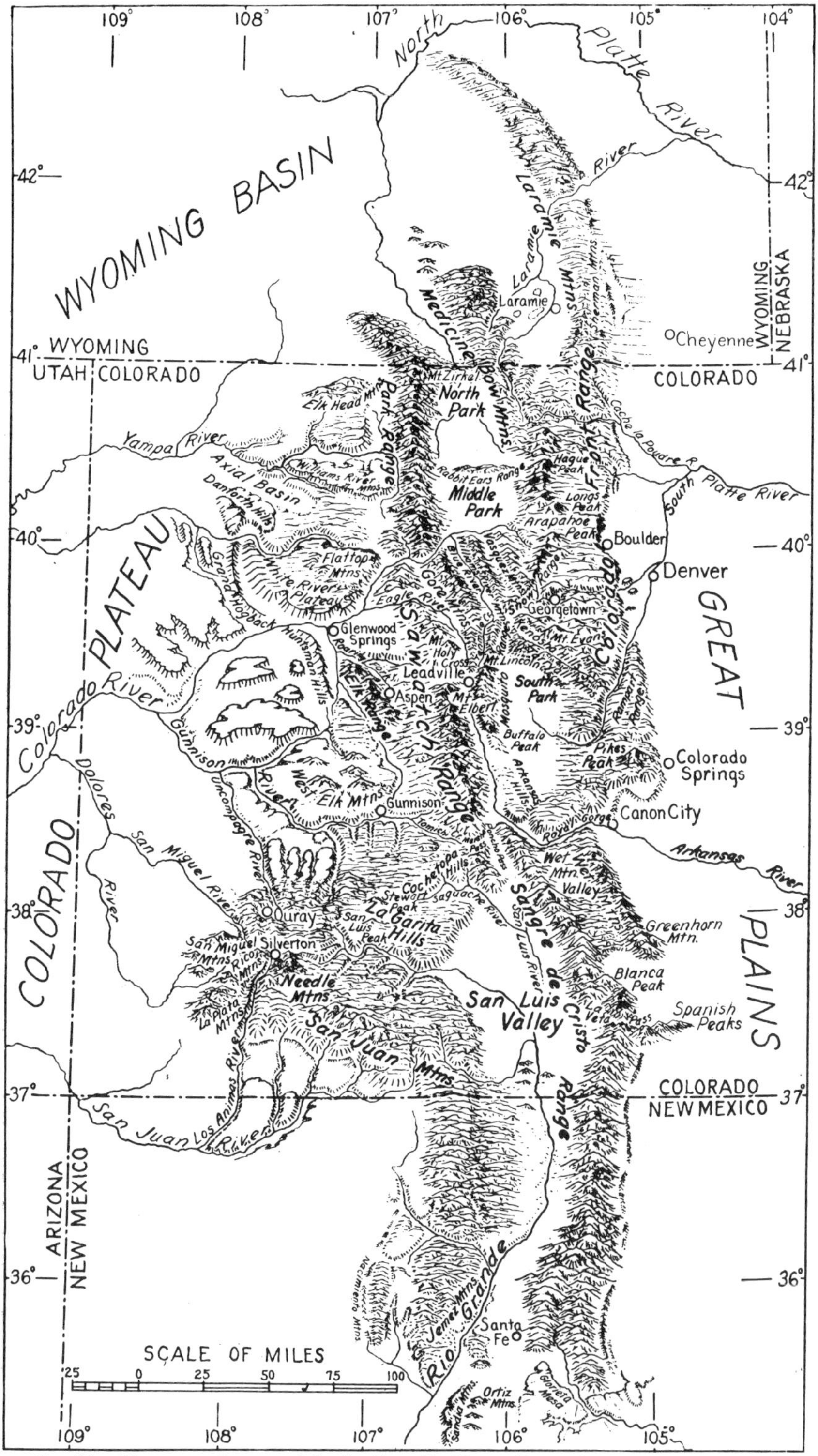

FIG. 98. Diagrammatic map of the Southern Rockies. (Drawn by Guy-Harold Smith for Fenneman: Physiography of the Western United States, McGraw-Hill Book Co., Inc.)

has the aspect of a dissected plateau. The present features are due largely to the glaciation of the summit area. Great cirques have been cut out of the upland and the remnants of the old pre-glacial surface are scalloped in outline. For a long distance the air route runs parallel with this range but not quite near enough to permit a thoroughly satisfactory view of these exceptionally interesting details.

The Great Basin. The Great Basin (Fig. 99), lying between the Wasatch and the Sierra Nevada, is distinguished by isolated, roughly parallel mountain ranges separated by desert basins which are almost level. The region is extremely arid, with the result that streams do not reach the sea and usually wither away in the separate

FIG. 99. Diagrammatic map of the Great Basin. (Drawn by Guy-Harold Smith for Fenneman: Physiography of the Western United States, McGraw-Hill Book Co., Inc.)

basins. The most important river of the region is the Humboldt. It drains most of northern Nevada before coming to an end in Carson Lake, which has no outlet. This lake also receives the water from Carson River draining the foothills of the Sierra Nevada. Truckee River, draining Lake Tahoe, ends in Pyramid Lake which does not discharge any of its water. Great Salt Lake, like the other lakes mentioned, has no outlet. This is true also of Sevier Lake. In all, there are several dozen separate basins, many of them containing salt lakes.

Numerous basin ranges are seen by the air traveller who crosses this province. The ranges vary in length from a few miles to 50 miles or more and are fairly straight and continuous. They are all maturely dissected, and have jagged and serrate crests. The mountain fronts are in many cases very abrupt, suggesting a fault origin. Long alluvial slopes, sometimes distinctly fan shaped, extend from the mountain foot into the basins. Some of these fans are deeply dissected with sharply cut ravines. The rocks which compose the basin ranges are in large part sedimentaries. From the airplane, because of the paucity of vegetation, the actual structure can in many cases be easily noted. Many of the ranges in the western part of the province include volcanic rocks in their make-up. Cinder cones and hot springs also occur. In one or two cases indications of recent faulting may be seen at the base of the ranges from the airplanes flying over them.

Much of the Utah section of the Great Basin Province, totaling about 20,000 square miles, was, during Quaternary time, covered by the waters of Lake Bonneville, the ranges then standing up as islands, just as Antelope Island does now in Great Salt Lake. In western Nevada a similar lake covered over 8,000 square miles of the basin area. The shore features of these two lakes, the terraces, beaches, bars, deltas, spits and other details, are beautifully preserved, largely because of the long-continued aridity of the climate. The beds of these former lakes are now vast plains of salt such as the Great Salt Lake Desert and Carson Desert, both of which are crossed by the air route. During times of heavy rain in the mountains the broad basin floors become covered with shallow playa lakes which soon disappear into the sand or evaporate. The strong winds blowing over these plains carry vast quantities of dust, producing wind-blown hollows or bolsons. These dust storms are frequently seen from the airplanes but always from a height far above them.

The Pacific Ranges. The Pacific Range Province (Fig. 100) which more or less coincides with the state of California in width, embraces three belts, namely, the Sierra Nevada, the Great California Valley, and the belt of the Coast Ranges.

The Sierra Nevada may be described as a gigantic tilted block mountain, its steep side facing eastward toward the Great Basin, and its upper surface sloping westward toward the Great California Valley. In the region around Lake Tahoe and Reno, where it is crossed by the air route, the Sierra Nevada consists of three ranges rather than one, each having the essential features of the single great range farther south.

Most of the rocks of the Sierra Nevada are granitic. In the central California portion lying under the air route, Tertiary lavas and tuffs were extruded near the crest of the range. These flowed westward in shallow valleys and thus displaced the streams. Later these displaced streams cut deep canyons between the tongues of lava, leaving the latter as high divides. Thus the lava now caps long ridges of upland, transverse to the trend of the range, and forms the divides. The upland surface or back slope of the Sierra is therefore partly a peneplain, and partly a surface determined by lava flows. Above this upland surface rise isolated monadnocks, whose summits represent a still earlier peneplain. In the central and western part they are not numerous, but they increase in frequency toward the main crest. West of Lake Tahoe the summits over a considerable area rise to this upper and older plane now profoundly dissected. The streams flowing westward down the back slope of the Sierra block are deeply intrenched, in some cases the canyons being 3,000 to 4,000 and even 5,000 feet deep.

The topography of the higher parts of the Sierra Nevada is largely the result of glacial sculpture. Most of the hills are rounded knobs or domes of granite, many of them without soil. Among these the streams pick their way, interrupted by lakes and falls. Some of the valleys are typical glacial troughs heading in cirques. With

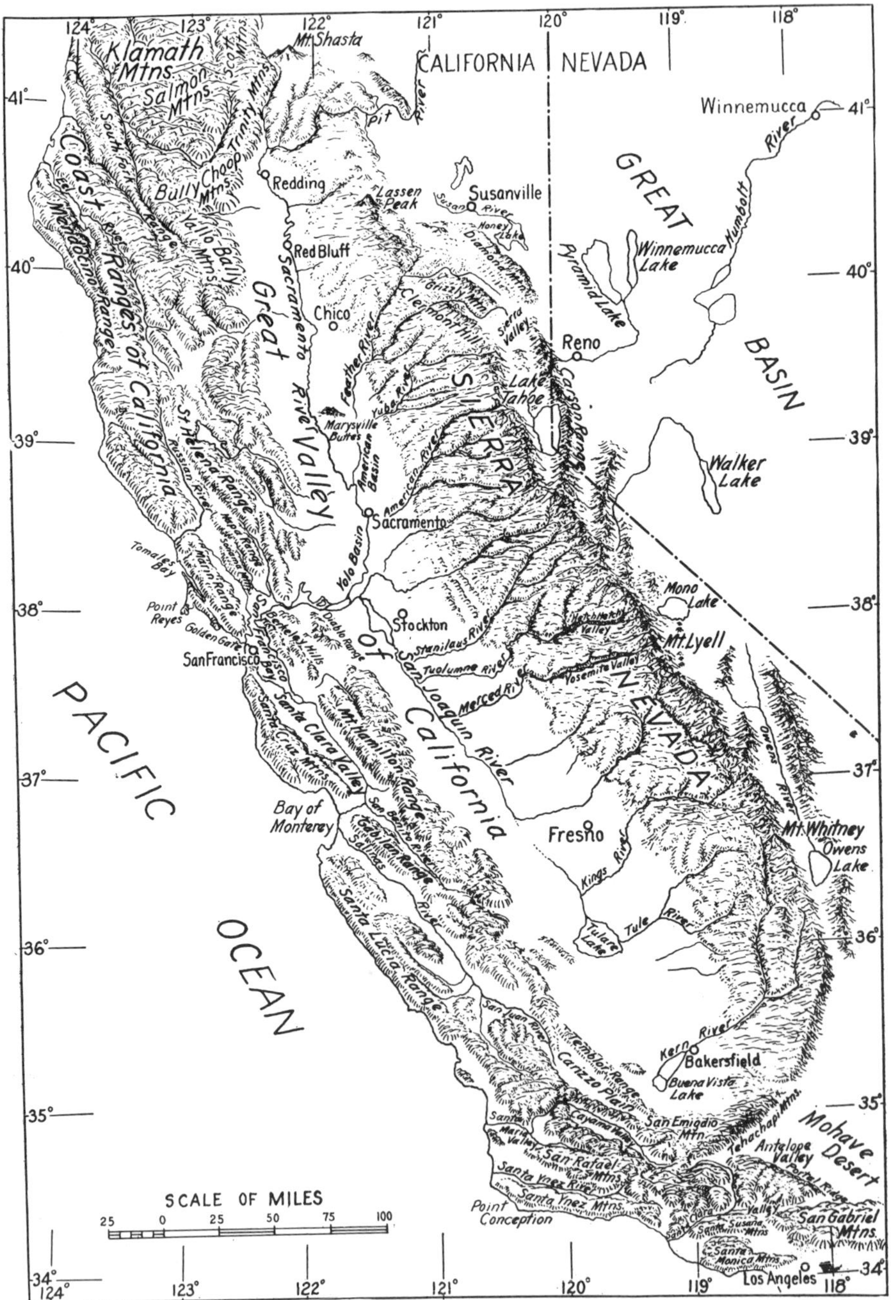

FIG. 100. Diagrammatic map of the Pacific Ranges in California. (Drawn by Guy-Harold Smith for Fenneman: Physiography of the Western United States, McGraw-Hill Book Co., Inc.)

lower altitude this ice-made topography of a recent epoch gives way to the more orderly branching valleys carved by the streams that now occupy them.

The Great California Valley is more than 400 miles long, with an average width of about 50 miles and an area of 20,000 square miles. Around it the mountains or their foothills rise abruptly, their boundaries being but little complicated by spurs and outlying hills. The larger part of the valley would be described by the casual observer as a dead flat. Considerable areas near San Francisco Bay are actually below sea-level, being protected from overflow by dikes and natural levees. Between the levee slopes of neighboring streams are basins, some of which are inundated or swampy.

The northern part of the California Valley is drained by the Sacramento River, the southern part by the San Joaquin River, both uniting in the so-called "delta" region at the head of San Francisco Bay.

An exceptional feature near the middle of Sacramento Valley is the group of hills known as the Marysville Buttes. These buttes are the remains of a nearly circular volcano about 10 miles in diameter. Their summits still rise 2,000 feet above the flat valley floor and form a landmark easily seen from the air route.

The Coast Ranges with their included valleys comprise a belt nearly 400 miles long with an average width of at least 50 miles. They consist of nearly parallel ranges of low mountains rarely over 4,000 feet high, with intervening valleys of various widths. The ranges do not run strictly parallel with the coast, but trend more strongly toward the northwest so that they abut against the coast, and the longitudinal valleys are open to the sea at their northwest ends. The longitudinal valleys are due not only to belts of weaker rock but even more to faulting, some of which has been so pronounced and so recent as to control the drainage and the major topographic features. The best known fault is the San Andreas rift made familiar by the San Francisco earthquake in 1906 which was due to a slipping along this line.

The submergence of some of the longitudinal valleys is responsible for the present San Francisco Bay.

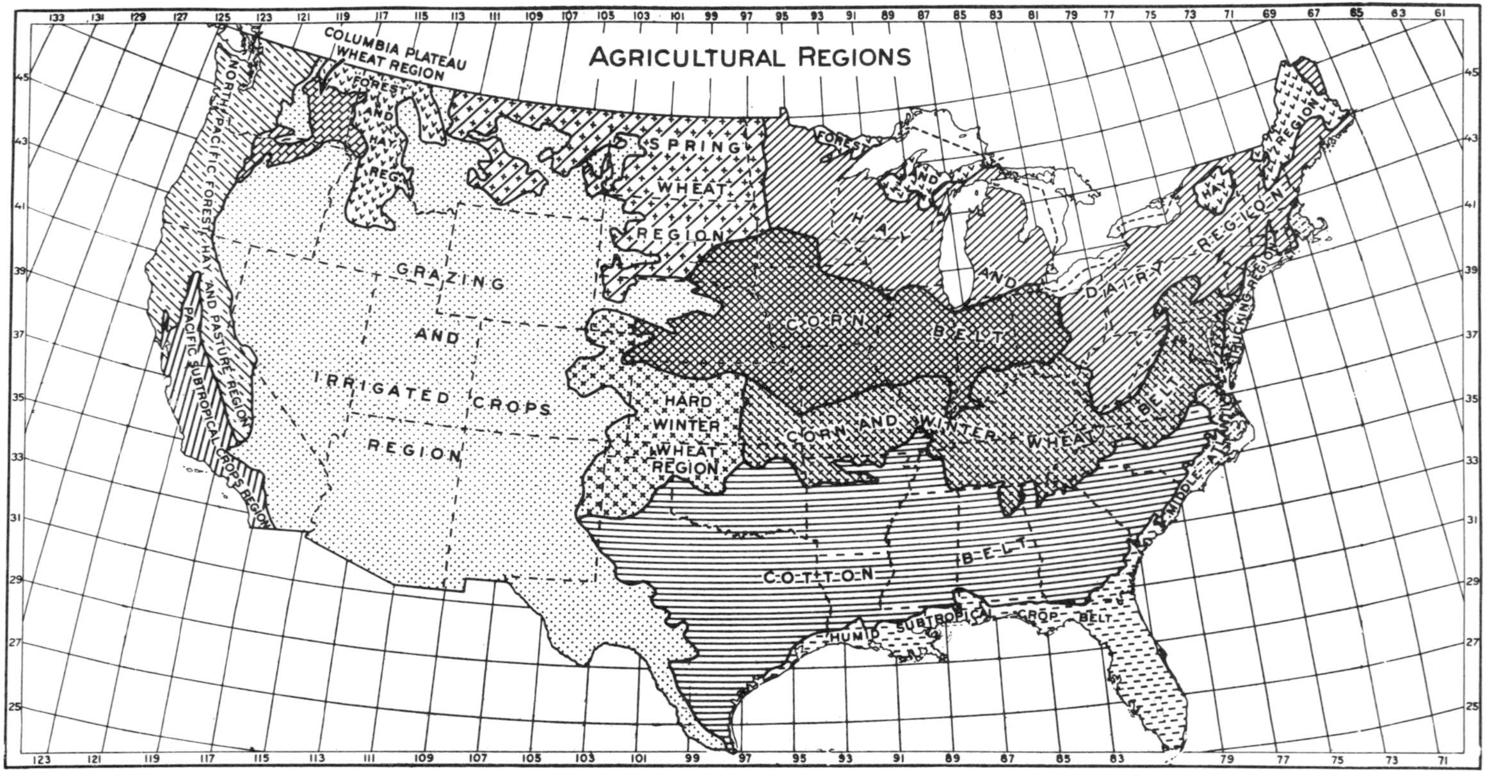

FIG. 101. A generalized map showing the major Agricultural Regions of the United States. (U. S. Dept. of Agriculture.)

II. THE AGRICULTURAL AREAS OF THE UNITED STATES

The United States may be divided into a number of agricultural areas (Plate III and Fig. 101) in each of which the type of farming is more or less the same throughout. Climate and soil characteristics figure largely in this division, and although in many instances the boundaries coincide with those of physiographic features, in others they do not. In some instances climatic or soil characteristics transcend in agricultural importance the physiographic features. A specific example is in the case of Glacial Lake Maumee, at the western end of Lake Erie, where the black soils extend somewhat beyond the limits of the old lake bed.

The air traveller is likely to spend more of his time observing the agricultural activities of the country than any other single set of items. A strip across the United States, along the air route, is therefore shown on Plate III with the different agricultural provinces outlined. This is part of a map now under preparation by Mr. C. P. Barnes, of the Bureau of Agricultural Economics of the Department of Agriculture. Through the courtesy of his Division it is reproduced here, prior to publication by the Department of Agriculture. The names used for the different agricultural units are those which have been adopted by that department.

The following description of the several agricultural areas is designed to present those aspects of each region which are most likely to be of interest to the air traveller during his journey across the country.

1. **Northern New Jersey Piedmont.** This is the glaciated portion of the Triassic Lowland. Its rolling uplands with reddish brown moderately stony soils are interspersed with many bodies of flat, poorly drained land. The several trap ridges, notably the First and Second Watchung Mountains, and Long Hill, are largely wooded. The region as a whole contains many towns suburban to New York City and much of it is used for residential purposes. There is also considerable market gardening, trucking and dairying.

2. **Northern Piedmont Upland.** The nonglaciated portion of the Triassic Lowland is sometimes classed with the Piedmont area. This lies south cf the glaciated portion and consists of reddish brown soils. There is dairying and poultry raising, and some general farming, devoted to corn, small grains and hay.

3. **Southern New England Uplands.** This belt is strongly rolling and hilly. It is represented by the New Jersey Highlands which are a continuation of the New England area. The portion crossed by the air route is nonglaciated, but soils and relief are not unlike those of the glaciated portion, lying farther north. There are light brown loams, but some of the soil is very stony. Dairying is the principal industry. Corn, oats, wheat and hay constitute the chief crops.

4. **Central Appalachian Valleys.** This region is part of the Folded Appalachian Province. Belts of gently rolling, productive, limestone-derived soils, alternate with belts of more strongly rolling, less productive soils derived from shales. The Central Pennsylvania Section crossed by the air route exhibits several pronounced ridges bearing a forest cover. In the valleys there is general farming with some dairying. Corn, wheat and hay are the principal crops.

5. **Central Appalachian Ridges.** This is the western part of the Folded Appalachian Province, in which the mountain ridges constitute a large portion of the area and alternate with valleys containing limestone-derived soil. There is general farming with some dairying. Hay, corn and small grains are the chief crops. The forested mountain ridges are essentially non-agricultural.

6. **Catskill-Pocono Highlands.** The Pocono section of this province has stony, shallow soils and very little agricultural development. It is mostly covered with forest or brush. There is some recreational development. Part of it is in State Forest. It is not actually crossed by the air route.

7. **Allegheny Plateau.** This is a high sub-maturely dissected plateau with steep slopes and relatively strong relief. Many rolling undissected remnants of the Plateau remain. A large part is too rough for cultivation by ordinary means. The soils are light colored and low to fair in productivity. Many are shallow and stony. Probably three-fourths of the region is in forest and cut-over forest land. General farm crops are grown, chiefly hay, oats, some potatoes and corn, and there is a little dairying. Agriculture is relatively unimportant.

8. **Upper Ohio Hills.** This is a maturely dissected, nonglaciated part of the Appalachian Plateau. It consists mainly of rolling hills with some steep slopes. The soils are partly limestone derivatives, light-colored, and of various textures. Hay, oats, wheat, corn and apples are grown and there is much pasture land, and dairying. Sheep and beef cattle are also important.

9. **Northwestern Appalachian Plateau.** The northwestern or glaciated part of the Appalachian Plateau is broadly rolling to undulating. The soils are mainly silt loams with compact subsoils and

poor underdrainage, of fair productivity when tile drained. Belts of morainic hills near the southeastern margin are well drained and more productive. Most of the region is in farms. Dairying is the outstanding industry. Hay and pasture land abound. Winter wheat, corn, oats and buckwheat are the chief crops.

10. **Erie-Ontario Lowland.** This province is restricted to a narrow strip of undulating country bordering Lake Erie. The soils are diverse. Heavy soils with heavy subsoils and poor underdrainage are extensive but there are also lighter soils. Proximity to the lake influences the climate and is responsible for fruit growing, particularly the cultivation of grapes. Dairying and truck-gardening are the principal agricultural industries.

11. **Maumee Blacklands.** In this level or slightly undulating plain with very dark, heavy, poorly-drained, but productive soils, corn, winter wheat, oats, hay and sugar beets constitute the chief crops. Dairying and swine raising are also important.

12. **Ohio-Indiana Silty Drift Plains.** This is a region of undulating or gently rolling lands with moderately light-colored productive, silty soils, interspersed with smaller bodies of level, dark-colored poorly drained silty or clay soils. The chief crops are corn, oats, wheat, hay and pasture, with swine raising, dairying and beef cattle as less important adjuncts.

13. **Northern Indiana Loamy Drift Plains.** These drift plains consist mainly of rolling land with light-colored, poorly drained loams and some muck. Corn, hay, pasture, wheat and oats are produced, as well as specialized truck crops on the muck lands. There is also dairying as well as the raising of swine, sheep and beef cattle.

14. **Kankakee Basin.** This region is generally level, with dark-colored, poorly drained sandy soils and muck, interspersed with some light-colored, sandy ridges. Here are to be found corn, pasture, oats, hay, wheat and rye; with specialized truck crops, particularly peppermint and onions on the muck lands. Beef cattle, hogs and dairying are also important items.

15. **Chicago Lowland.** This belt of dark-colored, poorly drained soils, sand ridges, dunes and sloughs is used almost entirely for urban and industrial purposes.

16. **Illinois-Indiana Prairies.** In this fairly large area the land is nearly level or gently rolling. The soil is dark brown, silty and productive. On the flatter areas the drainage is poor. Corn and oats are the cash crops but near Chicago dairying is important.

17. **Iowan Drift Prairies.** This land is undulating to gently rolling, with dark brown, productive loams and silt loams. The soils are generally somewhat lighter than in Area 18, next to be described, and are mainly well drained. Corn, oats, hay, and pasture are the chief crops with swine, beef cattle and dairying the chief animal activities.

18. **Iowa-Illinois Rolling Loess Prairies.** On these gently rolling lands the soils are very dark brown, productive, well drained and silty. The hills along the Mississippi have light-colored forest soils and are an exception to the general character of the area. As usual corn, oats, hay and pasture abound as well as swine and beef cattle.

19. **Upper Mississippi Hills.** This is a region of rolling hills with light-colored, silty forest soils, interspersed with smaller bodies of smooth dark-colored prairie soils. A relatively large part of the land is in pasture. Oats, hay, and corn are the most important cultivated crops with beef cattle the chief farm animal.

20. **Minnesota-Iowa Glacial Black Prairies.** These lands are level to gently rolling. The soils are very dark and productive. A large part of the area is drained. It is somewhat similar to Area 16. Corn and oats are the cash crops, with the raising of swine and beef cattle the important animal industries.

21. **Missouri-Iowa Drift Hills and Loess Flats.** In this area the lands near the streams are moderately to strongly rolling, but near the watersheds it is nearly level or undulating. The soil of the rolling lands is lighter in texture, less productive and easily eroded. Corn, pasture, oats and hay, with swine and beef cattle are the outstanding agricultural products.

22. **Missouri Valley Loess Hills and Rolling Prairies.** These are moderately rolling lands with broad rounded smoothly-sloping hills. The soils are very dark brown, productive, well drained, calcareous and silty. Corn, oats, hay, pasture, swine, and beef cattle are the usual products of the farms. There is some cash corn.

23. **Nemaha Drift Hills and Loess Flats.** Most of this region is moderately sloping, with dark brown, silty soils. Interspersed with these areas are nearly level tracts of dark brown soils with impervious subsoils, and a few steeply sloping lands with soils of lighter color and texture. The chief farm products are corn, some oats, wheat, hay and pasture, with swine, and beef cattle. There is also some dairying.

24. **Nebraska Plain.** This region is nearly level or slightly undulating. The soil is very dark brown and silty with a tough subsoil. The rainfall here is less than in the areas to the east. The heavy clay subsoil results in somewhat droughty soils. Corn and winter wheat are the chief crops. Some dairying is carried on, but swine and beef cattle are more important.

25. **Platte Sandy Plain.** This region coincides with the terraces and bottoms of the Platte River. The land is level to undulating with diverse soils, the larger part of which are sandy loams. Some bodies of heavier soils and some bodies of loose sand are included. Hay, particularly alfalfa, and corn are the outstanding crops. The fattening of sheep from western ranges is an important activity, and as usual there is the raising of swine and beef cattle.

26. **Republican Dissected Plains.** This is an undulating area with very dark brown, silty, productive soils, and somewhat calcareous subsoils. It is broken by belts of steep, eroded lands bordering the streams. The moisture is slightly below optimum for most crops, and yields vary markedly with the rainfall. Corn, wheat and alfalfa are grown, with swine and beef cattle the chief animals.

27. **Central Nebraska Loess Hills.** This region is strongly rolling with rounded, billowy hills, and some steep slopes. The soil is gray or grayish brown, calcareous, moderately productive and silty. It is suited to corn, alfalfa, other kinds of hay, and pasture. Beef cattle and swine as usual.

28. Great Plains Sand Hills. This unique area consists of rounded, billowy, steeply sloping hills of loose sand, supporting native grass cover. It is used largely for pasture and wild hay.

29. Northern Sub-humid High Plains. Like most of the plains this is a nearly level or undulating region with some dissection along the streams. The soils are dark brown with calcareous subsoils, and the low rainfall requires methods for conserving soil moisture. The crops are mainly corn and wheat, with a little pasture. Some beef cattle and hogs are produced.

30. Northern Semi-arid High Plains. The surface and soils of this region are similar to those of the preceding area but there is less rainfall, and as a result the dry-farming land is of a lower grade. It is devoted mostly to winter wheat and pasture, with some cattle.

31. Platte Piedmont. This piedmont basin is a region of diverse topography and soils. Bodies of undulating, rolling land alternate with hills and ridges. The soils are both heavy and light. The rainfall is low and dry farming is generally successful only in years of abundant rainfall. Grazing of live stock is the principal use of the land, but there are a number of small irrigated valleys and some dry-farmed lands producing forage crops.

32. Southern Rocky Mountains. The section crossed by the air route consists mainly of the Laramie Mountains, a range of rounded hills covered with a thin growth of yellow pine and used chiefly for grazing. The Medicine Bow and Elk Mountains, which are steep, rocky and largely forested, may be viewed at a distance.

33. Outer Wyoming Basin. This area of arid range land is in part smooth, as for example the Laramie Plains, which are crossed just west of the Laramie Mountains. Most of it, however, is rolling or broken. It is generally sage-covered, although the Laramie Plains support short grass. As a whole it is only fair to poor grazing land.

34. Red Desert. The very dry central portion of the Wyoming Basin is known as the Red Desert. It has generally less than 10 inches of rainfall annually. The land is in large part alkaline, with much shadscale and affords poor grazing.

35. Wasatch-Uinta Mountains. The Wasatch Mountain Range, crossed by the air route is rough, broken and in large part forested. There is some grazing in the valleys.

36. Utah Piedmont Valleys. The nearly level or gently rolling valley lands at the foot of the Wasatch Range have brown soils of wide textural variety. The average annual rainfall is between 15 and 18 inches, permitting dry-farming, but irrigation is practiced where feasible. Sugar beets and alfalfa are the outstanding crops. There is, however, a wide variety of other crops, including fruits and specialized truck and there is also some dairying.

37. Outer Great Basin. The outer sage-covered portion of the Great Basin, although arid, is moister and cooler than the intermontane deserts, next to be described. It has an average annual rainfall of more than 10 inches, and consists of fair to poor grazing land, interspersed with numerous mountain ranges that are better watered and bear semi-arid woodland.

38. Intermontane Semi-Deserts. These dry, shadscale-covered basins, generally with an average annual rainfall of less than 10 inches have alkaline soils. They alternate with mountain ranges bearing semi-arid woodland. Poor grazing land makes up most of the area but there is also some irrigated land producing mainly forage crops.

39. Intermontane Alkali Deserts. The salt flats representing old lake beds contain scattered growths of salt desert shrubs. The region is essentially an absolute desert with little or no grazing or other use.

40. Sierra Nevada. This high and rugged mountain area consists mainly of steep and rocky lands, with occasionally some soil smooth and deep enough for cultivation. The rainfall and snowfall are both heavy. Lumbering and grazing are the principal surface uses of the land.

41. Sierra Nevada Foothills. The lower western slope of the Sierra Nevada consists of rolling lands interrupted by steep and rocky areas. The soils are of various textures and some are shallow and stony. The lower portions of the region have a very long frost-free season because of favorable air drainage. Fruit growing is the outstanding enterprise. Both deciduous and citrus fruits and nuts are produced. There is also grazing of sheep and cattle.

42. Sacramento Valley. This portion of the great California Valley is nearly level with some rolling lands near its margins. There is a great diversity in its soils. Some are hardpan, some are poorly drained, and some are muck. The area is mainly sub-humid with most of the rain coming in the winter. Crops are grown both with and without irrigation. There is a wide variety of crops, with a tendency to specialization in different localities. Fruit, particularly grapes, peaches and prunes, as well as nuts, potatoes (on the muck of the "Delta" of the Sacramento and San Joaquin Rivers), rice, field beans, wheat, barley and alfalfa all find a place here. Sheep and lambs constitute the chief farm animals.

43. San Joaquin Valley. This region is somewhat drier than the Sacramento Valley. Among the locally specialized crops are fruit, especially grapes and peaches, as well as nuts, vegetables, field beans, alfalfa and barley. Sheep and lambs are raised and dairying is active.

44. Central California Coast Ranges and Valleys. This is a region of chaparral and grass-covered mountain ranges, interspersed with semi-arid valleys having nearly level floors. The summers are dry. Fruit, especially prunes and apricots, beans, vegetables for seed, sugar beets, and small grains cut for hay constitute the chief crops. Beef cattle are produced and dairying is important.

45. Bay Coastal Belt. This littoral belt, famous for its dense summer fog, has cool, almost rainless summers, mild winters, and a long frost-free season. The hills are interspersed with nearly level lands. There are some extensive tidal marshes. The crops are specialized, and agriculture is of the intensive type with fruit growing, trucking and dairying important. The portion seen from the air route is largely in urban development.

III. THE NATURAL VEGETATION REGIONS OF THE UNITED STATES

Second only in interest to the air traveller, after the agricultural aspects of the country, is the natural vegetation, transcending even the geology in its apparent importance. The following brief notes have been compiled from the splendid atlas on Natural Vegetation by H. L. Shantz and Raphael Zon issued by the Department of Agriculture. At the outset it will be understood that the original natural vegetation of the country has over much of its area been replaced by agricultural crops, orchards, pastures, and man-made forests. But this will only serve to emphasize the relationship which exists between the natural vegetation and the various agricultural or forest crops which supplant it.

The vegetation of the United States may be broadly divided into forest, grassland, and desert shrub. (Plate IV.)

Forests

The forest vegetation forms two broad belts, one extending inland from the Atlantic Ocean and the other inland from the Pacific Ocean. The eastern is relatively continuous, while the western is broken by many interspersed areas of non-forested land. Part of the western forest is represented in the Rocky Mountains and is detached from the rest of the western forest area, which lies closer to the Pacific coast. It therefore divides the non-forested portion of the United States into two broad strips, one of which lies east of the Rocky Mountains and the other between the Rockies and the Sierra. Of these two non-forested belts the eastern constitutes the great grassland area, while the western constitutes the desert shrub area.

The eastern or Atlantic forest is essentially broad-leaved in composition and of unbroken distribution, covering both valleys and mountains while the western forest, which is distinctly coniferous, is characterized by abrupt changes in type, and is interspersed by treeless valleys.

The eastern forest was formerly unbroken but now only one-tenth of its original area remains and even less than that proportion in the belt covered by the air route.

The western forest still corresponds largely with its original natural limits, although large parts have already been made unproductive by unrestricted lumbering and by destructive fires which have swept over enormous tracts. Two-thirds of the forest area of the United States is concentrated in the eastern half of the continent. The remainder is on the western side and is mostly coextensive with the Rockies and the Cascade-Sierra and Coast Ranges.

Grassland

Lying between the western and eastern forest belts and extending from Canada on the north to Mexico on the south, is the great grassland area, broken only by river courses and occasional buttes or low mountains. Grasses of one kind or another grow in this region because they are admirably suited to withstand conditions of excess drought and fires which would destroy tree growth. Such grasslands usually are well supplied with water in the surface soil during the growth period and do not depend to any extent upon deeply-stored soil moisture. Grasses are characteristic of regions of summer rainfall. A great profusion of showy flowering plants is usually found within the grasslands, and they often present a varied appearance during the early part of the season. About 70% of the grassland east of the 100th meridian is now under cultivation, but in the west only about 10% is used in that way.

Desert Shrub

Between the Rocky Mountains on the east and the Sierra on the west lies the great inland desert, characterized largely by zerophytic shrubs, plants able to live under extremely dry conditions. Here and there the desert area is interrupted by intrusions of grasslands or at higher elevations by forests. This whole area is characterized by a deficient rainfall and an excessive evaporation rate. The perennial vegetation which gives character to the deserts consists principally of shrubs. The relatively small size of the plants together with their small leaves, well protected against excessive transpiration, the small amount of growth produced each year, and the wide spacing, enable these plants to conserve the scanty moisture supply and continue their growth during the long rainless periods. Such dry periods may occur at any time of the year. Of the desert-shrub area about 2% is under irrigation.

THE EASTERN FOREST AREAS

In the eastern part of the United States crossed by the air route two types of forest may be distinguished: one, the Northeastern Hardwood Forest; and two, the Southern Hardwood Forest.

The Northeastern Hardwood Forest. This belt covers much of New England and southern New York and extends across the plateau section of western Pennsylvania, where many square miles of it may be seen almost unbroken and practically in its virgin condition. The characteristic trees of this belt are birch, beech, maple and hemlock, basswood and elm. In the higher and more rugged parts of the plateau the hemlock and other coniferous trees are more abundant than elsewhere. The air observer will readily note that the densest stands of this forest are along the sides of the deep valleys which indent the plateau, the upland surface in many places having been cleared for the production of such crops as timothy, clover, oats, barley, corn and buckwheat.

The Southern Hardwood Forest. The lower slopes in the Appalachian region and the central Mississippi Valley support a hardwood forest in which the oaks make up the great body of the forest. Three types of this forest may be distinguished. In the more rugged sections of Pennsylvania lying on either side of the high Allegheny Plateau together with the contiguous parts of adjacent states is a forest region where chestnut, chestnut oak and yellow poplar abound. The region is essentially mild in its winter and has everywhere a fairly high annual precipitation running to as much as 70 inches.

West of this belt, in Ohio, Indiana and Illinois, is the oak-hickory forest which borders the prairie areas and extends along the water courses. It is this forest which has been largely cleared away in the central states with only small patches remaining now as wood lots on the farms.

The third type, represented by the oak-pine forest of the Coastal Plain and Piedmont area, is not traversed by the air route.

THE GRASSLAND AREAS

Tall Grass or Prairie Grassland. The original prairie vegetation which greeted the eyes of the first emigrants to the west had the appearance during the spring of a veritable flower garden, composed of phloxes, shooting stars, violets, spiderworts, and other showy plants, almost to the exclusion of the more important but slower growing and less highly colored grasses. This original vegetation has now been almost entirely replaced by cultivated crops. As the season advanced the area was dominated by tall, luxuriant and relatively deep-rooted grasses. In this region the rainfall is heaviest during spring and summer. But in the fall the moisture supply within reach of the grass roots is often entirely exhausted by the luxuriant growth of grasses, and droughts occur. During drought periods the area has in the past been repeatedly burned by fires started either by Indians, travellers, or lightning, making it impossible for trees and shrubs to survive and thus giving the area over to grasses which are able to maintain themselves under such conditions. Since the settlement of these lands and the consequent checking of the prairie fires, tree growth has been gradually extended either by planting or natural seeding, and trees now grow throughout the whole prairie region, except in the driest wind-swept situations.

Within the area covered by this type of vegetation lies the most valuable body of agricultural land in the United States. Much of the Corn Belt is included in this area, but is by no means confined to it. Winter wheat is a most important crop in the west central portion of the prairie region. Oats and hay are also produced in large quantities throughout the prairie region.

Short Grass or Plains Grassland. Short grass characterizes the Great Plains lying east of the Rocky Mountains and west of the 100th meridian. The grasses are low-growing and shallow rooted, owing to a low precipitation which does not penetrate more than two feet before it is exhausted by the grass cover. During dry years this great expanse presents the appearance of an endless carpet while in wet years taller plants develop on the short grass sod and give to it a more mixed appearance.

Dry farming has been undertaken in nearly all parts of the short grass region, but it has proved profitable only in the less arid portions. Under irrigation the soil generally is productive and has been most successfully employed in growing cereals, alfalfa, sugar beets, potatoes, and other vegetables. Although most of the land has been taken up, less than one-eighth was in harvested crops in 1919. In the east-central portion corn and winter wheat are important crops.

Bunch Grassland. This grassland is characteristic of much of the Pacific coast states. In Oregon, Washington, and California it occupies extensive areas, the largest being the Great Valley of California. Throughout the northern Great Basin this bunch grass type occurs at higher elevations, below the conifer zone. It is characterized by a rich display of flowering plants. The condition under which bunch grass develops is that of a moisture supply insufficient for a dense stand of grasses. The rainfall varies from 10 to 25 inches. Some of the moisture supply comes during the growing season. If rainfall were confined to the growing period short grasses would develop, whereas if a still greater proportion fell during the winter rest period, it would give way to sage-brush. All of the area is valuable grazing land and in California especially the bunch grasses have long since disappeared as a result of overgrazing, and only the weed grasses remain.

DESERT SHRUB AREAS

The desert shrub area lying between the Rocky Mountains and the Sierra supports two types of vegetation, namely, one: the sagebrush or northern desert shrub type; and two: the greasewood or salt desert shrub type.

The Sagebrush Region. This region is characterized by a scattered open stand of deciduous shrubs, almost all of which have small leaves of a light or silvery color, presenting a very monotonous appearance. The sagebrush regions comprise a much greater area than any of the other desert types. The rainfall of much of this region is less than 10 inches per year, much of it coming during the winter rest period and finding its way deep into the soil. Many of the characteristic plants are deep rooted. The sagebrush plant grows from 2 to 7 feet high and resembles a small tree. The plants always grow several feet apart, a stand of sagebrush therefore resembling an open miniature forest. During the spring and early summer the leaves are light green or silvery in color but they are gradually shed in the fall and the bare stems give a dark tone to the landscape. The roots of this plant extend to depths as great as 18 feet. Sagebrush covers the plateau country of southern Wyoming and the alluvial fans of the Great Basin. Its use for grazing purposes depends upon the small amount of grass and other herbage which develops on the interspaces.

The Greasewood Region. The Great Salt Lake Desert, the Carson Desert, and other salt flats of the Great Basin support clumps of shrubs which are able to withstand appreciable amounts of salt in the ground water. Regions covered by greasewood and other similar plants usually present the appearance of vast white plains of salt-encrusted soil with clumps of plants widely scattered over the surface. The soil between the plants is covered with little or no perennial growth. The plants are green in color, due to the succulent leaves, and when in full leaf present a relatively luxuriant appearance, contrasting sharply with the gray of the shadscale or sagebrush. This contrast in color is readily noted by the traveller crossing in spring or early summer from the desert shrub part of the Great Salt Lake Desert to the sagebrush regions elsewhere in the Great Basin.

THE WESTERN FOREST AREAS

The air traveller will probably be able to distinguish only two or three of the several types which make up the western forest, namely the western pine forest region characterized by yellow pine and Douglas fir; the chaparral, or region of scattered and usually small broad-leaved trees; and the piñon-juniper regions.

The Western Pine Forest. The western pine forest covers the Rocky Mountains, the Wasatch, the Sierra Nevada and parts of the Coast Ranges. It includes several different kinds of pines as well as the Douglas fir. The yellow pine occurs on dry, hot slopes such as the western slope of the Sierra, and in the Rockies where the rainfall is between 20 and 30 inches a year. It is more or less characteristic of the transitional zone between the grassland and the forest areas. It does not form a continuous forest, but often includes open grazing lands that give a park-like effect to the region. This effect is heightened by the comparative absence of underbrush and the presence of a fairly dense growth of grass and other herbaceous plants beneath the trees. The trees themselves are usually large, growing in groups, or widely spaced, so that the sunlight reaches the ground in all parts of the forest with but little interference by the crowns. Open park-like forests of this type may be noted on the Laramie Range and especially on the eastern slope of the Wasatch Mountains.

At higher elevations, especially in the Rocky Mountains and in the higher parts of the Sierra upland, Douglas fir occurs mixed with the yellow pine.

Chaparral. Chaparral is a mixed forest of stunted hardwood trees and shrubs. It occupies a belt below the yellow pine and above the desert shrubs. This type of vegetation occurs throughout the foothills of the central Rockies but is most typical of the Coast Ranges of southern California. Along the lower slopes of the Sierra and on the Berkeley Hills the traveller may observe this kind of forest. Numerous species of plants make up the chaparral, the oak being one of the most common. The trees are in many cases low and shrub-like, sometimes standing in open groves but often forming an impenetrable thicket. Most of the chaparral region is not well suited to agriculture, because of the low rainfall and the steep slopes which render irrigation water inaccessible.

Piñon-Juniper Forest. This belt, like the chaparral, lies below the yellow pine zone and above the zone of desert shrubs. Most of the ranges of the Great Basin are clothed in this type of vegetation. Over thousands of square miles piñon-juniper and sagebrush alternate, the former occupying rough broken country or shallow stony soil, while sagebrush occurs on the more level ground, which has a deep uniform soil.

The area of land occupied by piñon-juniper, especially in the Great Basin, is very great. Economically these trees are important, since they form the chief source of timber for mine props, fence posts and fuel for local use.

IV. CLIMATIC AND SEASONAL ASPECTS OF THE AIR ROUTE

Precipitation. On the basis of total precipitation, the United States may be divided into an eastern and a western part (Plate V). The dividing line follows approximately the 100th meridian, which runs through central Texas, Kansas and Nebraska, in the vicinity of which the average annual precipitation is about 20 inches. In general, the eastern part of the country has sufficient precipitation for successful farming by ordinary methods, but in the western part the amount of moisture over large areas is insufficient for the requirements of crop growth and special methods of supplying moisture to the soil or of conserving it are often employed.

In the Pacific coast states there are great differences in precipitation within short distances. On the Sierra Nevada and on the Coast Ranges of California the annual precipitation is over 80 inches but in the Great California Valley there are only 15 and even less than 10 inches in many places; and in the western part of the Great Basin at the foot of the Sierra the rainfall is less than 5 inches per year.

East of the Rocky Mountains precipitation is comparatively uniform over large areas. Between the Atlantic Ocean and the Mississippi River in the region traversed by the air route there is a variation of only 10 inches a year between the places having the greatest and those having the least rainfall.

On the Pacific Coast winter is the wet season. Near San Francisco and on the Sierra Nevada, over half the precipitation of the year comes in December, January and February. During the warm summer months, from April to September, at San Francisco only about 3 inches of rain fall, practically none of this coming in June, July and August.

On the basis of the annual amount of precipitation as well as of its seasonal distribution, four precipitation types may be distinguished in going across the United States. From east to west, they are: first, the Eastern type; second, the Plains type; third, Great Basin type; and fourth, the Pacific type.

In the *Eastern type,* which includes the region lying east of the Mississippi River, there is a comparatively uniform distribution of precipitation throughout the year, with possibly a slight decrease in the autumn months. The rainfall in this region is due to the regular procession of cyclonic storms which cross this part of the country and are best developed during the winter period. As summer advances thunderstorms become common and supply much of the rainfall of that season. Corn thrives under these conditions.

The *Plains type* covers the region between the Rocky Mountains and the Mississippi River. It is characterized by generous rains in the late spring and summer months and very light late fall and winter precipitation. Most of the summer rainfall is due to thunderstorms which are more abundant over this region than elsewhere in the United States. Forty to sixty thunderstorms occur each year in most localities.

The total rainfall for much of the plains area is under 20 inches per year. Much of this falls in late spring and early summer just at the time when it is needed for plant growth. Cheyenne, for example, has 88 days each year with precipitation, but twice as many days of rainfall in each of the summer months as in the winter months. Vegetation in this belt is tufted grass, the closeness of the individual tufts varying with the amount of precipitation. Sagebrush and similar drought-resistant shrubs are common on the drier margins. Grazing is one of the most important industries.

The *Great Basin type* is characterized by very scanty precipitation, usually less than 10 inches and in some parts less than 5. Relatively large seasonal and diurnal ranges of temperature are characteristic. The arid heat of summer and dry cold of winter are less hard to bear than similar temperatures in more humid climates. Rock disintegration, due to wide variations of temperature between day and night and between winter and summer, is rapid with the result that many mountains of the Great Basin appear to be buried under their own waste. Sagebrush and other drought resistant shrubs, widely spaced, are the characteristic vegetation. Bare ground is everywhere in evidence. The vegetation is capable of supporting only a very sparse animal life. Only at oases spots, usually along mountain-fed streams, like at Reno, Nevada, and at Salt Lake City, are there large clusters of population.

The *Pacific type* has a marked winter concentration of precipitation and a summer dryness. This type is sometimes called the Mediterranean type because it is so characteristic of the Mediterranean region. Cool winters and long hot summers, with large diurnal ranges of temperature are the rule. The dry atmosphere and clear skies are conducive to hot days and cool nights. The almost entire lack of summer rain encourages such industries as fruit drying in the California Valley. Along the coast fogs of marine origin are common summer-time phenomena, causing the nights and early mornings to be damp and unpleasant, but by ten o'clock the mists have been evaporated by the ascending sun.

In both the Pacific and Great Basin areas, including the Rocky Mountains, the mountain masses exert a marked influence on the precipitation, not only over the mountains themselves but often over large areas to the leeward of each range. From the Rocky Mountains westward the distribution of precipitation is largely determined by topographic conditions. The surface of this region is characterized by a succession of mountain chains, either in massive ranges or partially detached groups of peaks, with intervening narrow valleys, usually trending north and south. This disposition of the land relief results in large variation in the amounts of precipitation, even in near-by localities.

Winds. The importance of the wind as a factor in air travel may be realized from the fact that the actual scheduled flying time from the Pacific Coast to the Atlantic Coast is three hours less than the scheduled flying time going west. Only between Omaha and Chicago is there no essential difference in the eastbound and westbound schedules. This is due to the fact that Chicago is at the end of the long run from the Pacific Coast and the schedule for the last lap of this run is less rigid, so that planes can arrive on time even under adverse conditions en route. Between New York and Chicago, for example, the eastward flight is made in one hour and twenty minutes less time than the westward flight. Between Omaha and Cheyenne the difference is almost an hour in favor of the eastbound schedule. Between Cheyenne and Salt Lake City, and also between Salt Lake City and San Francisco, there is in each case an advantage of one hour in flying time on the eastward journey in contrast with the westward. (These were the 1932 schedules.)

The prevailing westerly winds blowing across the country have an average velocity near the surface of the earth of between 5 and 15 miles an hour in different parts of the country, being greatest in the Great Plains and least in the basin areas between the mountain ranges of Wyoming and of central California. It is a climatic fact of great economic importance that the wind velocities over the Great Plains are so extraordinarily adapted for driving windmills. The windmill is, therefore, one of the most characteristic features of the Great Plains. At higher altitudes the velocity is much greater, hence the practice of flying eastward at altitudes of 10,000 feet and westward at only 1,000 to 2,000 feet above the surface of the earth.

The contrast between eastward and westward flying may be best appreciated by considering the effect of only a 10-mile wind, during a flight of 5 hours. On the eastbound trip this means a tailwind which adds 50 miles to the distance travelled in 5 hours. On the westbound trip this is a headwind which reduces the distance travelled by that amount. The total difference between the two flights is 100 miles which is the best part of the distance flown in one hour. A wind of 20 miles per hour is not at all uncommon. This means almost two hours difference in flying time. Or, if we consider a run of 400 miles with a wind of 40 miles an hour, we find that it takes almost twice as long to travel this distance against the wind as with the wind. It is not difficult to understand, therefore, why air schedules can not be adhered to so rigidly as railroad schedules. The schedules are made to fit average conditions. It is no uncommon thing to leave Reno, Nevada, half an hour behind time and reach Salt Lake City, the next stop, over 400 miles to the east, ahead of schedule.

The strength and direction of the prevailing winds are not quite the same in summer and in winter. Local topography, especially in the mountainous part of the United States affects the wind directions. In general it may be said that during the winter the winds blow from the west or from the northwest, but in summer, especially east of the Rocky Mountains, southerly winds are, in general, of most frequent occurrence and of longest duration. Plate V shows by means of arrows the prevailing winter and summer winds.

Airplanes head into the wind when they land and when they take off. When surface winds on the landing field average more than 20 miles per hour, flying conditions are apt to be unsatisfactory, chiefly because of gustiness. With winds of this average speed the instantaneous maximum values may reach 30 miles per hour falling off almost as quickly to 10 miles per hour. It is these rapid changes in speed, especially when accompanied by rapid fluctuations in direction, as is often the case, which sometimes make the ground handling of aircraft difficult. If there is no gustiness, a moderate wind is an advantage since it gives a lower ground speed for the same air speed, when heading into the wind. Thus in close quarters a plane can get off or land with a shorter run.

Cyclonic Storms. Although practically the whole of the United States lies within the region of the "westerlies," common to all middle latitudes, the weather is largely controlled by the movements of areas of low and high barometric pressure and the attendant characteristic winds peculiar to each. The weather conditions which accompany a typical winter storm are represented on the following diagram (Fig. 102) which is simply an idealized storm so commonly shown on the so-called synoptic weather map of the U. S. Weather Bureau.

As the storm center or region of low pressure advances across the country, travelling approximately 500 to 1,000 miles a day, the region in front of it experiences a change in wind direction and temperature. The winds blow toward the center of the storm. East of the storm front, therefore, the winds blow from easterly points, being northeasters in the northeastern quadrant and southeasters in the southeastern quadrant. In the latter case, because of the southerly winds, the temperature usually rises. At the same time the skies become overcast and there may be precipitation, either rain or snow, depending upon the period of the year. After the storm has passed the wind comes from the west or northwest, occasionally from the southwest. The temperature usually drops, the barometer rises, and there is a period of cool clear weather until the next low approaches.

Before starting on any flight the pilot familiarizes himself with the synoptic map of the area. Figure 103 shows the map of a storm that made flying difficult east of the Allegheny Mountains during the month of March, 1930. Figure 104 is the weather report issued to the pilots flying between New York and Chicago on this same date. It gives the weather conditions for 18 different points between those terminal cities, that is about every 40 to 50 miles.

Each terminal airport weather station such as Newark, Cleveland, Chicago, Omaha, Cheyenne, Salt Lake City and Oakland receives the regular weather information that is exchanged four times each day between the many weather stations in the United States. From these reports the familiar synoptic weather map is drawn, indicating the general position of the barometric high-pressure and low-pressure

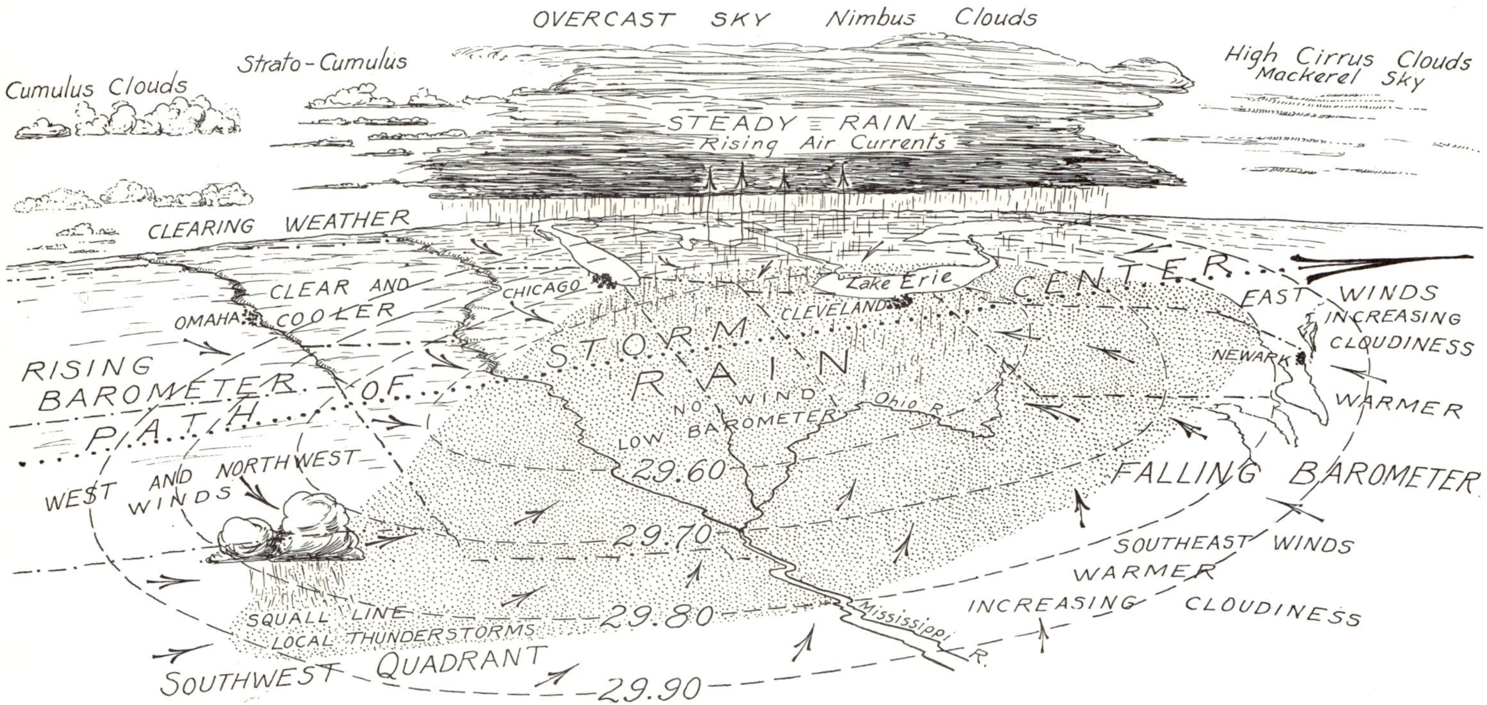

FIG. 102. Diagrammatic representation of a cyclonic storm passing over the eastern United States. The storm moves eastward. The wind comes from all directions toward the center of the storm. In the eastern and southeastern quadrants (e.g. at Newark) the wind is from the east or southeast, the barometer is falling, the temperature is rising and a mackerel sky betokens a coming storm. If in summer, the weather may be oppressive and muggy. On the western side of the storm, at Omaha, the reverse conditions occur; the wind is from the west or northwest, the barometer is rising, and the weather is becoming clear and crisp.

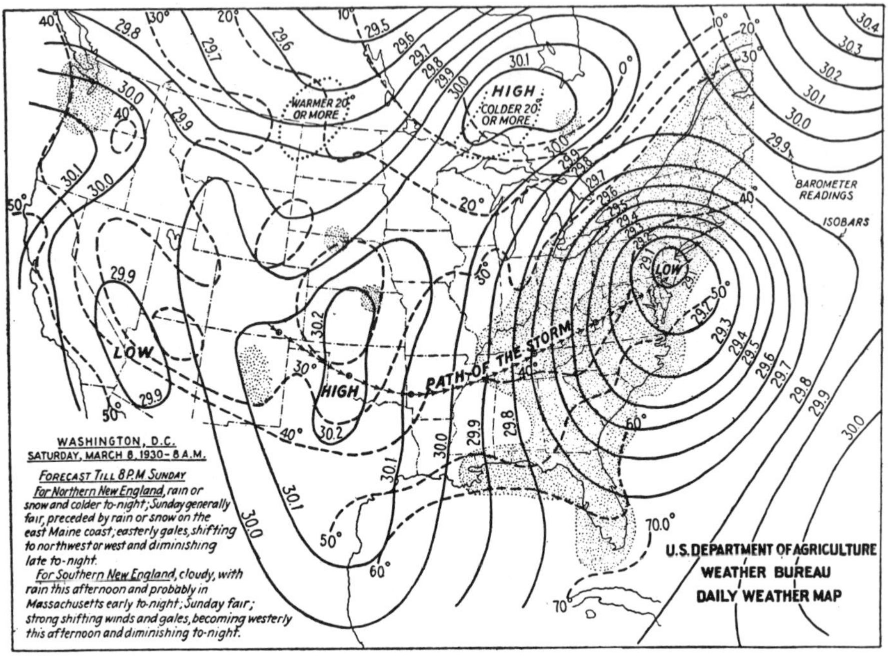

(From Smith: *Air Transport Operation.* McGraw-Hill Book Co.)

FIG. 103. Synoptic map of a storm that made flying difficult east of the Allegheny mountains.

U. S. DEPARTMENT OF AGRICULTURE
WEATHER BUREAU

PILOT'S WEATHER REPORT

Date March 8th, 1930. Time 8.00 A. M. Issuing station Cleveland, Ohio.

STATION	ELEVATION (FEET)	TIME (A. M. OR P. M.)	GENERAL CONDITION	CEILING (FEET)	VISIBILITY (MILES)	WIND DIRECTION AND VELOCITY	TEMPERATURE (DEG. F.)	DEW POINT (DEG. F.)	BAROMETER (INCHES)	FIELD CONDITION	REMARKS
Hadley		8-A.M.	O-C Light Fog	500	1/8	E NE 10	41	40	29.09		Misting
Allentown		"	O-C Light Rain	500-E	4	NE 10	41		29.04		Light Haze
Park Place		"	Dense Fog	none	none	NW 8	37		29.16		Sprinkling
Numidia		"	B-C Light Rain	600-E	[illegible]	NW 5	39		29.16		Moderate Fog
Sunbury		"	O-C	800-E	3	N 12	38		29.22	Soft	Light Haze
Winklebleck		"	Dense Fog	None	none	N NE 10	33		29.33		Few Snow Flakes
Bellefonte		"	O-C Misting	600	1	NE 4	35	34	29.22		Light Haze
Kylertown		"	O-C Misting	200	2	NW 6	34		29.32	Soft	
Greenwood		"	O-C Light Fog	300	2	NW 10	32		29.29	Soft	Misting
Brookville		"	O-C Lower B-C	1000-E	Unl	N 8	36		29.24		Misting
Mercer		"	O-C Light Haze	500-E	5	NW 15	34		29.46	Soft	
Parkman		"	O-C Light Fog	400-E	2	N 20	33		29.46		Light Rain
Cleveland		"	O-C Misting	600	3	N 17	34	33	29.49		
Vickery		"	O-C Light Fog	400	[illegible]	NW 28	33		29.58	Soft	Moderate Snow
Toledo		"	O-C Lower S-C	1200	4	N 28	34		29.70		Light Haze
Helmer		"	O-C	1500-E	6	N 10	30		29.80		
Goshen		"	O-C	2500-E	6	NW 15	29		29.89		
McCool		"	O-C	800	10	N 20	32		29.90		
Chicago		"	O-C	1200	5	N NW 14	28	21	29.92		

O-C: Overcast. E: Estimated Ceiling. B-C: Broken Clouds. S-C: Scattered Clouds. Unl: Unlimited Ceiling.

(From Smith: *Air Transport Operation.* McGraw-Hill Book Co.)

FIG. 104. New York-Chicago Airway weather report during the storm of March 8, 1930.

areas at that time. In addition to the above, weather reports are obtained at least hourly from selected points along each of the various airways, approximately 60 miles apart. The spacing and location of these airway reporting stations are determined by the nature of the country. Weather reports are obtained from the highest available points to determine whether they are enveloped in clouds. They are also obtained from some of the low points to have information regarding any ground fog that may exist, and likewise from the different airports along the route so that the pilot will know whether a landing is possible at those points, should it be necessary. Weather reports usually come from caretakers at the airway routing beacons at the high points and from the attendants at airports along the airway. All reports give: 1, the general weather conditions; 2, the ceiling or height of the lowest clouds; 3, the visibility along the ground measured in miles; 4, the wind direction and velocity; 5, temperature; 6, barometric pressure; and 7, other facts of general or unusual interest, such as surface conditions on the field.

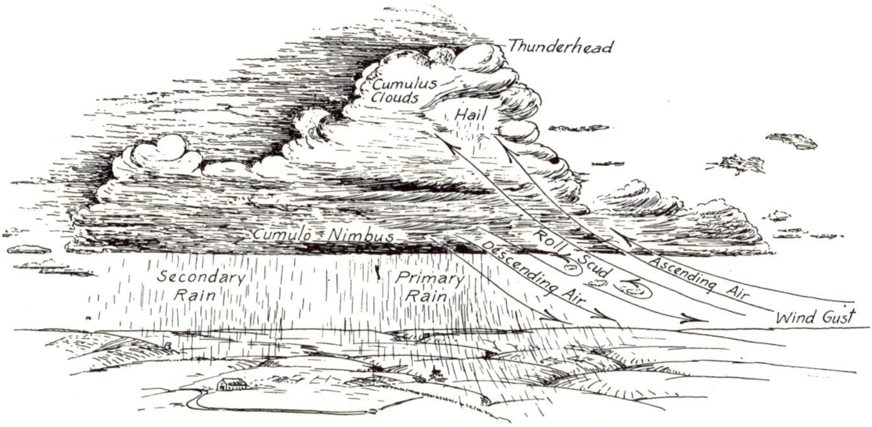

FIG. 105. Diagrammatic representation of a thunderstorm. A thunderstorm seems to advance against the wind because the air in front is being sucked into the storm and descends again in the rear. The rapidly ascending air moving toward the front of the storm produces the "wind gust," which blows dust about and heralds the immediate approach of the storm. The "roll scud" consists of small detached clouds driven rapidly along under the mass of the storm just preceding the rain. Hail is a common feature of thunderstorms because of the great height to which the air currents rise, probably about two miles. Most thunderstorms cover a relatively small area of country and are of short duration.

Barometric Pressure. The difference in barometric pressure between places several hundred miles apart renders the altimeter untrustworthy unless proper corrections can constantly be made during actual flight. A change of one inch in barometric pressure is equivalent to almost 1,000 feet difference in elevation. Changes of half this amount are not unusual on an airplane flight of over 400 miles. Without being aware of barometric changes due to the weather, it is obvious that a pilot could under those conditions not know his correct altitude within several hundred feet. However, the radio telephone stations regularly advise all pilots of changes in barometric pressure, enabling them to make proper corrections and thus accurately to determine their altitude at all times.

In this connection it is also interesting to observe that the landing speed of planes is greatly increased in regions of high altitude, where the air is more rarified. At Cheyenne, for example, a landing speed as well as a taking-off speed of almost 80 miles an hour is necessary for planes which at Chicago require a speed of only 50 to 60 miles an hour.

Thunderstorms. Most of the largest and best-developed thunderstorms of the eastern United States occur in or close to the transition zone between the warm, muggy weather brought by the southerly winds in front of a passing cyclonic area, and the cooler, drier, clear weather which comes with the westerly and northwesterly winds in its rear. (See Figure 102.) This is usually in the southwestern quadrant of the storm, and is known as the wind-shift or squall line. It is due to the fact that the winds from the north and west moving toward the center of the storm swing rather far to the south and, because of their coolness and consequently greater weight, burrow under the warmer and lighter winds coming from the south. Along this wind-shift line there often develops a row of thunderstorms progressing eastward as a body. Because of the large area involved in such thunderstorms a pilot has little chance of flying around them, whereas in the case of the more localized thunderstorms of summer it is a comparatively simple matter to fly around the disturbance.

A cross-section of a typical thunderstorm is shown in Figure 105. The great anvil head of a cumulo-nimbus cloud rises to heights of 30,000 to 40,000 feet above

(Photo by U. S. Weather Bureau)

FIG. 106. Fog, coming in from the Pacific Ocean, through the Golden Gate, San Francisco. The Presidio of San Francisco is shown in the foreground.

the earth's surface. With the approach of the storm the stillness which usually precedes it gives way to a sudden gust of wind blowing toward the storm. The danger to aeroplanes of a storm of this kind is due not only to the great violence of the winds but to the strong upward and strong downward currents which prevail, as well as to the fact that they change with great suddenness. The extent of the weather reporting service, however, makes it possible for pilots to be warned by radio telephone of the approach of storms into which they should not fly.

Much has been said and written relative to the danger to airplanes from lightning. Regular operation has proven that there is very little danger of an airplane being struck by lightning while it is in the air. The hazard involved is mostly a mental one to the pilot caused by his observations of damage that lightning has done when it strikes the ground. Lightning may occur in a local thunderstorm, in a line squall or wind-shift line, or in a tornado or hurricane. When any of these sorts of storms are reported along an airway, the meteorologist should be able to determine from the general weather conditions and from the specific weather reports just which of these types it is, and to advise and warn the pilots accordingly. This is particularly necessary for night flying where the pilot can see only the lightning that accompanies the storm.

In the winter time, when the temperatures are low, line squalls usually consist of much vertical turbulence and heavy snow squalls, and the pilot must be advised and warned of these possibilities by a message from the ground radio station. Such matters as air temperatures at the different altitudes, wind direction in and above the clouds, the thickness of the clouds, the presence or absence of cloud layers, and any region where ice may tend to be precipitated upon the airplane are facts readily ascertained by pilots during their trips and passed on to the meteorologist who in

(Photo by U. S. Weather Bureau)

FIG. 107. Stratus clouds, at two different levels. The lower one, practically at the surface of the ground, consists of long wisps of clouds. The upper layer is relatively high and might be called alto-stratus.

turn uses these data to inform others following the same route. It is not hard to understand from this that an experienced pilot becomes weather-wise and puts to immediate and practical use all of the meteorological facts which can be assembled for his benefit.

At some time in the future it may be possible to fly planes on schedule, regardless of weather conditions, by the aid of better airplanes and engines, better instruments, radio beacons, radio landing beams, and full and complete protection of the airplane from the ice hazard. But even so, the pilot will always have to know all that is available about the current weather conditions and the immediate changes that may occur in order to make his flight in safety and on schedule. The pilots of transport planes receive such frequent and accurate information prior to taking off, and over

FIG. 108. Clearing after a storm in central Pennsylvania. Alto-stratus clouds at a very high level, with cumulus clouds just below the plane.

the radio telephone when in flight, that they are able to avoid flying into bad weather, which they could not always avoid prior to the adoption of two-way radio telephone and the perfection of the weather reporting services on the air lines.

Clouds. The air traveller, during the course of his trip, sees several kinds of clouds, some of which add to the glorious beauty of the sky through which he flies; others only delay his journey.

Fog may be considered as a type of low cloud. As distinguished from other forms of atmospheric obscurity, such as mist and haze, a fog is practically the same

FIG. 109. Nimbus or rain clouds covering the Uinta Mountains. The plane is flying eastward in the rain but the sun is breaking through the western sky and a rainbow appears, just beyond the wing of the plane.

(*Photo by U. S. Weather Bureau*)

FIG. 110. Nimbus, or rain clouds, with fog or stratus below. Low nimbus clouds covering the mountain summits make flying hazardous, but high nimbus clouds over flat country are not serious.

(*Photo by U. S. Weather Bureau*)

FIG. 111. Cumulus clouds over the Great Plains. These are fair weather clouds. On the Great Plains, because of the curvature of the earth, the most distant clouds reveal only their tufted summits

density throughout. Viewed from a distance a fog has a definite boundary. The fogs which roll in almost every afternoon through the Golden Gate (Fig. 106) and over the California Coast Ranges are certain to be seen by the traveller in that region. Heavy fogs occur around Cleveland south of Lake Erie, in the mountains of central Pennsylvania, and around New York Harbor. In all of these places they are a real obstacle to aviation because of the delay occasioned.

A *stratus* cloud (Fig. 107) is a horizontal sheet of lifted fog. It may be very low, only a few feet above the surface of the ground, in which case there is said to be a very low ceiling or it may hang 2,000 feet high or even more. When that is the case airplanes readily fly under it. When stratus clouds rise to heights of 6,000 to 30,000

(Photo by U. S. Weather Bureau)

Fig. 112. Cumulo-nimbus clouds, with probably some rain falling beneath, producing a local thunderstorm.

or even 40,000 feet above the earth's surface, they are called alto-stratus (Fig. 108). On such days the sky is said to be overcast though, if thin, the sun may be seen dimly outlined as through ground glass. Such clouds give no difficulty to air navigation unless they happen to rest upon mountain tops over which the flight is made.

A *nimbus,* or rain cloud (Fig. 109), is a thick layer of dark cloud, without shape and with ragged edges, from which continued rain or snow generally falls. Through openings in these clouds an upper layer of high stratus may almost invariably be seen. These clouds are usually several thousand feet above the earth's surface and are the type which accompany cyclonic storms. Flying under such conditions is not unduly hazardous unless the rain or snow is very heavy (Fig. 110).

Cumulus clouds or wool-pack clouds (Fig. 111) look like big tufts of cotton in the sky. The base of these clouds is horizontal but the upper surface is dome-shaped and exhibits protuberances. These clouds often bedeck the summer sky. As their average height is 4,000 feet, much flying is done above them and through them. These clouds are formed by an ascensional movement which is almost always noticeable so that aeroplanes flying beneath them are apt to meet bumpy conditions.

A *cumulo-nimbus* cloud (Fig. 112) is a thunder-cloud or shower-cloud. These are heavy masses of clouds, rising in the form of mountains, turrets, or anvils. From the base, local showers of rain or snow usually fall. Clouds of this type should not be approached by aircraft as they indicate violent vertical air motion. Vertical velocities as high as 2,500 feet or more per minute may be experienced in the front of a thunderstorm cloud.

A *cirrus* cloud (Fig. 113) has a delicate and fibrous appearance often showing a featherlike structure, generally of a whitish color. Cirrus clouds appear sometimes

(Photo by U. S. Weather Bureau)

FIG. 113. Cirrus clouds, tufted form, at high altitude, with stratus clouds in the distance.

as thin filaments on blue sky arrayed in parallel belts which by an effect of perspective appear to converge toward a point on the horizon. These clouds are extremely high, rarely below 15,000 feet and usually much higher, far higher than aeroplanes ever go. At such high altitudes these clouds are normally composed of ice crystals. When cirrus clouds assume the form of flakes or globular masses, more or less regularly arranged, they produce the well-known mackerel sky.

Distance of Visibility from Different Heights. From an elevation of 100 feet above level ground objects on the ground 13 miles away are visible. Such a view encompasses something over 500 square miles. In the middle west this equals 15 ordinary townships.

From the low flying elevation of 1,000 feet above the earth's surface, which is common for west-bound flights, the view extends 41 miles in all directions and encompasses over 5,000 square miles of country.

From an elevation of 10,000 feet which is frequently the height of the plane above the ground, the country is visible for 132 miles, representing in all an area of over 50,000 square miles, more than that of New York State.

In the far west, mountain peaks and ranges, especially if snow capped, are readily seen 100 miles distant and under good but not uncommon conditions for 150 miles. During the early morning and late afternoon hours the sun is visible from the plane while the earth beneath is enshrouded in darkness and shadows. From a height of 10,000 feet the sunrise may be seen more than 10 minutes earlier than from a point on the ground.

Length of Day and Night Along the Air Route. The most northern point on the air route is Chicago, on latitude 42 north. The most southern is San Francisco, approximately on 38 north. Forty degrees north may be taken as an average for the whole route. At this latitude the sun rises at 4:30 on June 21, the longest day of the year, and sets shortly after 7:30. Twilight in June extends for two hours before sunrise and after sunset. During the winter, on December 21, the sun rises about 7:15 and sets shortly after 4:30 with twilight lasting only an hour and a half in the morning and evening. This means that the summer day begins to break about 2:30 A. M., but the winter day not until almost six o'clock. Night-flying passengers will be most interested in these facts. Darkness does not come until 9:30 at night during the summer but in winter it gets dark shortly after six. This means that the winter day has about 12½ hours of light, at the most, whereas in summer, daylight extends over more than 19 hours.

On September 21 and March 21, the sun rises and sets about six o'clock with twilight periods of less than an hour and a half.

V. AIRWAYS AND AVIGATION

History and Development of the Air Route. The route followed by the United Air Lines from New York to San Francisco was the original transcontinental air mail route, first operated by the U. S. Post Office Department in 1920. The Department of Commerce and United Air Lines have completely equipped this route with radio stations for communication between ground and planes, and the entire route has been equipped by the Department of Commerce with beacon lights, intermediate landing fileds and weather reporting service for day and night flying from the Atlantic to the Pacific coasts.

In the earlier days of air transportation it was assumed that an airplane could be flown between any two points without regard to the topography along the course, and, as a result, the original airway of the air mail route consisted only of landing fields at various regular stops. These landing fields were very primitive and the maps which the pilots had to use were extremely crude and unsatisfactory. If the pilot encountered either mechanical difficulties with his plane or bad weather, as he frequently did, he had to choose some farmer's pasture or grain field for his emergency landing and hope that there were no holes, stumps, or ditches hidden beneath the growth on the field.

This system of informal emergency fields continued until night flying was inaugurated. It then became necessary to provide lighted intermediate fields for safety. Night flying at first was done only by accident, when the pilot found himself overtaken by darkness on a late trip and had the alternative of a forced landing in some unsuitable field or of continuing his trip in a plane that was not equipped for night flying over an unlighted airway. There were no markings or large signs along any of the original airways by which a pilot might check his location. If he could not do this by the meager information on his map, he was usually compelled to fly low enough over a town to read its name from the sign on the railroad station.

When night flying was proposed for the air mail, the relatively flat Mississippi Valley between Chicago and Cheyenne was chosen as the logical place in which to start the development work. Emergency fields were established 25 miles apart and outlined with boundary lights. A revolving beacon was installed at each field. Between the emergency fields, spaced at intervals of three miles, were placed flashing gas beacons such as had been used to mark the intersections of highways.

The original lighted airway between Chicago and Cheyenne was established in 1923 and an experimental period of night flying on schedule was carried out in September of that year. Regular scheduled night flying was started over this airway as a part of the transcontinental air mail service on July 1, 1924. The lighted airway was extended eastward to Cleveland and westward to Rock Springs, Wyoming, in the fall of 1924 as the longer nights of the winter months made this necessary. From the results obtained over this stretch of lighted airway, it was decided to light the airway from New York to Cleveland so that the mail could be sent overnight by plane from New York to Chicago. This work was undertaken during the winter months of 1924 to 1925 and was ready so that the overnight service between New York and Chicago could start on July 1, 1925.

The problems involved in lighting mountainous country were very different from those encountered in lighting fairly flat country. The revolving beacons between New York and Cleveland were equipped with 1000-watt bulbs and 24-inch reflectors. These were mounted on the high points as well as on intermediate fields, so that their spacing averaged not more than 10 miles but in some cases where intermediate fields were adjacent to high mountain tops they were not more than 5 miles apart. Between New York and Cleveland there were only certain fields along the airway that could be used

Fig 114. The intermediate landing field at Shafter, Nevada. View taken flying westward from an elevation of about 2,000 feet above the ground.

safely by airplanes either by day or by night, because of the unevenness of the country. Almost every field along the airway that was known to be a good one was therefore lighted, with the result that the spacing between the intermediate fields varied from 5 to 30 miles, depending upon the contour of the country.

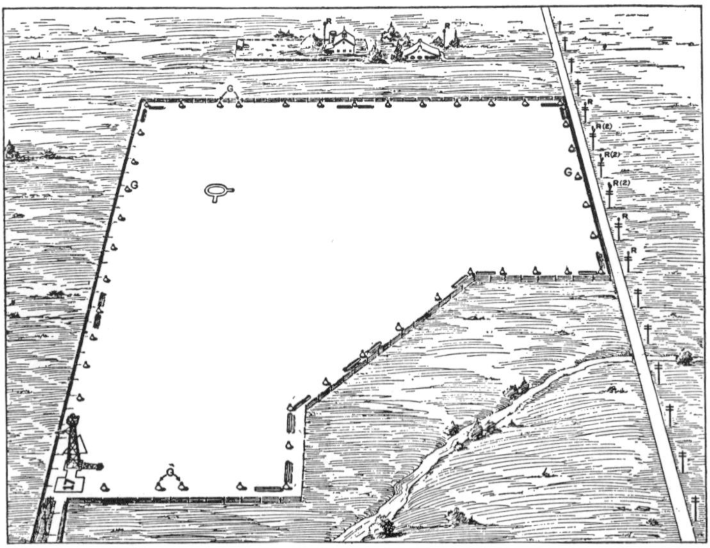

(*U. S. Dept. of Commerce*)

Fig. 115. Typical Department of Commerce intermediate landing field. Green lights indicating best available runways are shown at G and red obstruction light at R. R(2) indicates two red lights. A beacon light is shown at the lower left corner and boundary lights and markers are shown around edges of the field.

For the present, an airway may be considered to be a path through the air, approximately 10 miles wide, between two important commercial centers. For night flying, this path is outlined on the ground by rotating beacon lights of either 1,000,000 or 1,250,000 candle power spaced approximately 10–15 miles apart. At intervals of 40 to 50 miles there are located intermediate or emergency fields (Figs. 114 and 115). The routing beacons form a line as nearly straight as is possible, their real location being determined by the contour of the country. In mountainous country routing beacons are mounted upon the highest peaks along the way, to mark the lowest level at which an airplane may be safely flown. This means that in mountainous country beacons will probably be neither in a straight line nor accurately spaced.

FIG. 116. Boeing, twin-motored 10 passenger biplane, traveling at 160 miles per hour between New York and San Francisco.

The spacing of intermediate fields at approximately 40 to 50-mile intervals has been found to provide satisfactorily for interruptions of flights due to weather conditions. There are some who may think that in the interest of perfect safety fields should be spaced 10 miles or closer and that planes should be flown at such elevations that they could effect a landing by gliding at any time. This would necessitate flying at altitudes of 4,000 or 5,000 feet. At this altitude wind velocities up to 80 miles an hour are not uncommon in the eastern half of the United States, and over the Rockies they may go as high as 120 miles per hour. These strong winds are generally the west winds, prevailing in these latitudes. It would be quite impracticable to fly against winds of this velocity. As a result, westbound flights are sometimes made at altitudes of not over 1,000 or 2,000 feet above the highest points along the route, whereas eastbound flights are made at higher altitudes. It is practically impossible to have fields so close together that planes could always find one available for reach by gliding from these low elevations used on the west-bound flights.

Operation of the Air Lines. Air travel today is much more than just a plane in the sky, for back of the millions of dollars which have been invested in late type equipment, including improved, dependable power plants, is a far flung and extensive ground organization, comparable to those established by older forms of surface transportation. In flying over the 2776 mile airway from New York to California, via Cleveland, Toledo, Chicago, Des Moines, Omaha, Cheyenne and Salt Lake City, the passenger traverses the first transcontinental air line.

More passengers and mail are flown over the New York-Chicago-San Francisco route than over any long distance route in the world, and on it United Air Lines flies approximately 7,000,000 miles annually, of which more than half is at night. It is interesting to know that the New York-Chicago-San Francisco route is likewise the longest lighted airway in the world.

The public is showing increased knowledge of airplanes, motors, personnel and operating practices, but for those who are unfamiliar with the technique of commercial aviation, the following facts may be of interest.

The predominant type of equipment operated by United Air Lines on this route is a low-winged all-metal, Boeing monoplane (Fig. 116), powered with two 550 h.p. supercharged Wasp motors. These planes, cruising approximately 160 miles an hour, carry ten passengers, two pilots and 1,000 pounds of mail and express. The trend of airplane manufacture now is toward low-wing, monoplanes, as distinguished from the biplane of a few years ago. Furthermore, the trend is towards high speed, medium sized airplanes, rather than the thirty and forty-passenger planes talked about in recent years.

The Wright Brothers airplane which made the first successful flight, had a twelve horse-power motor, but the airplane engines have been developed until today some of them have a weight of only 1.2 pounds per h.p., a record achieved by the radial cooled, Hornet engine.

The newer airplanes are equipped with every proved aid to commercial aviation, including two-way radio, the reception apparatus for the directive radio beam which holds the pilot to his true course, even if visibility is obscured, a directional gyro for pointing the true course of flight, an artificial horizon, indicating the attitude of the plane in flight, together with the other instruments customarily installed on the modern transport.

Operation of Planes. The actual operation of a plane by the pilot may be best understood by noting the instrument panel (Fig. 117) of the new Boeing twin-Wasp speedy transports recently placed in service on the mid-continent and other routes of United Air Lines. This is perhaps the most complete instrument installation to be found in any commercial aircraft. In the cockpit of the ten-passenger and cargo plane there are, of course, numerous controls and switches which do not appear on the panel.

The first step in beginning a flight is the starting of the engines. These are warmed up for several moments before passengers and cargo are put aboard. (1) indicates the primer for the two supercharged Wasp engines, each of which develops 550 horse-power. When the engines have been properly warmed up, the pilot taxies the plane up to the passenger depot where passengers, mail, baggage and express are stowed aboard. Then the pilot runs the plane to the end of the runway, where he "revvs" each engine to make positive that the motors are functioning properly. Then he opens the throttle and takes off.

When the pilot's altimeter (13) registers 2,000 feet, or whatever altitude is most advantageous for the particular flight, he levels off and adjusts his course by checking the directional gyro (8) and his compass, not mounted on the instrument panel. He then sets the landing wheel retracting mechanism into operation, pulling the wheels up into the wings until the indicator (2) registers directly on "up." Above and to the right of this indicator is a red bull's-eye light which flashes warningly if the pilot starts to land without putting his landing gear back down to landing position. The light is supplemented by a klaxon horn so that no pilot can forget to adjust the landing gear prior to landing.

Instruments used by the pilot during flight are grouped strategically in the center of the panel. In addition to the directional gyro (8) and sensitive altimeter (13), these include the artificial horizon (9), air speed indicator (10), the turn and bank indicator (11), the rate of climb indicator (12) and the time of flight clock (14).

To the right of the center section are grouped the engine instruments and gauges. These include the fuel pressure, oil and temperature gauges (15, 16 & 17), the tachometers, indicating revolutions per minute for both engines (18), and the ammeter (20), showing if batteries are charging or discharging.

Other important instruments and gauges include the air temperature indicator (3), the engine cylinder temperature indicator (4), the engine cylinder temperature selector (5), the gasoline supply gauges (6 & 7) and the fire prevention device (19).

Also on the instrument panel are reserve radio fuses, contained in a box mounted above the ammeter. Every twenty minutes, the pilot or co-pilot of the plane reports to ground stations by radio telephone, advising the ground personnel of the position of the plane in flight, and in turn receiving latest reports on weather conditions. In the event a fuse in the radiophone apparatus should blow out, a very infrequent happening, it may be immediately replaced.

All United Air Lines pilots are trained to fly by instrument alone, and the group of instruments in the center of the panel, coupled with the radio telephone and the directive radio beacon service, make air navigation an exact science.

Pilot Personnel. The requirements for employment as a transport pilot on the large air lines are rigid. Not only must pilots have proven experience but they must maintain a high standard of personal conduct and physical condition. The average flying experience of the 150 United Air Lines' pilots is approximately 4,000 hours. Included among these fliers are nine, each of whom has flown in excess of 1,000,000 miles, and a substantial number of them have 10,000 hours to their credit.

Most of the veteran pilots gained their first experience during the war, but the younger ones, which includes men like Lindbergh, started their careers since 1918. Texas was one of the main training grounds for the United States Air Services during the World War, because flying conditions in winter were good, and because there is plenty of level land, which is highly desirable for student pilots.

The life records of pilots show that they gained their experience in many ways. One of the younger men began by barnstorming in Oklahoma. Then he drifted into motion pictures and flew as a pseudo air warrior in such plays as "Hazardous Valleys" and "Air Circus." Two of the older pilots, in the early days of aeronautics, conducted a flying circus up and down the Middle West.

Another one of the older pilots, since he commenced flying in 1917, has spent a twelfth of his life in the air. His imposing total of over 8,000 flying hours means an average of a solid month out of each year off the ground. Included in that 8,000-hour total are almost 3,000 hours of night flying.

Several of the younger men began as school teachers, and apparently their lives as pilots have been no more thrilling since they made the change. Many of the pilots with 5,000, 6,000, and 7,000 hours to their credit have never had an accident of any kind, and report that they have participated in no outstanding flights. "None" is

FIG. 117. Instrument panel of the twin-motored Boeing 10-passenger biplanes.

written on their reports after "Records held," "Achievements," and "Outstanding Fights." Two thousand flights over the Sierra Nevada from Reno to San Francisco, 500 trips over the Alleghenies between New York and Cleveland, hundreds of trips over the Great Plains from Chicago to Cheyenne, these are the modest records of the veterans who now pilot the big transport planes from coast to coast. Some of the pilots have flown over every division of the transcontinental route, but most of them have their own run between two division points.

Pilots as a rule are poor talkers. They would rather fly than talk. Nor do they get into the papers very often. One of the veterans of the air mail, however, appears regularly in the news because every once in a while he adds 100 to the number of times he has flown over the Transcontinental "Hump" between Reno and San Francisco Bay. In 2,000 trips over the Sierra, with air mail and passengers, he has had only one forced landing and that without serious accident. Each day he flies the air mail between Reno and San Francisco in 2½ hours, a ten-hour trip by train. When he is not flying a big tri-motored 12-passenger plane, he is piloting one of the high speed mail-cargo ships carrying some 60,000 letters.

From piloting, some of these men move to administrative positions in the company's service where their long experience and training is invaluable. They know in a most intimate way the conditions which exist over the lines they are directing.

Ground Organization. Four people are required on the ground for each plane aloft, and as the modern airplane speeds swiftly across the skyways, credit should be given to the skilled, efficient, painstaking operating and mechanical personnel stationed at frequent intervals along the various routes.

The pilot personnel of United Air Lines numbers 150 and there are approximately 500 in the ground employee group, foremen, engine mechanics, helpers, inspectors, electricians, radio operators, dispatchers, instrument repair men, wood and wing workers, welders, painters and clerks. The close coordination of activity between the air and the ground crews is attested to by the unusual and dependable operating record.

The plane is ready and waiting all serviced when the pilot arrives a half hour before leaving time. Planes are flown approximately 700 to 1,000 miles on each run.

The inspection which precedes flight is highly systematized. There are nineteen checks to be made in engine inspection, nine for the fuselage, seven for the wings, besides those for landing gear, lighting system, controls, service and adjustments. It is policy to leave nothing to chance and there are complete engine overhauls every 200 hours. Propellers are etched for possible flaws every 200 hours.

Due to the length of its routes, the company has built a veritable chain of depots and servicing plants, this investment running into hundreds of thousands of dollars.

In addition to thirteen permanent airports at which the coast-to-coast planes stop, there are emergency fields spaced approximately thirty miles apart and numerous municipal and private fields.

Weather Reports. The United States Weather Bureau has a very extensive and nationwide weather reporting service for airplanes. Forty-five upper air meteorological stations have been established by the Weather Bureau. They furnish information at intervals, night and day, to the weather control stations along the airways. Reports are also received from 200 first order weather bureau stations twice daily. These, in combination with the upper air reports (Fig. 118), are used in making weather forecasts, available to all pilots using the airways.

Teletype circuits for transmission of hourly reports between weather reporting stations have been established on routes flown by United Air Lines. At certain auxiliary fields, spaced about fifty miles apart, there are airway weather observers with teletype communication with other stations. Weather reports are collected and broadcast from all airway radio stations by radiophone. This broadcast is made at least every hour so that pilots may have first-hand information regarding flying conditions along the route and at terminals.

Prior to taking off, pilots have the latest weather information (Fig. 119). As they fly over the airway they receive over the radiophone: first, spot weather information from places on the airway, and second, local and general forecasts of the weather conditions in each of their zones. This can be supplemented by information phoned

(*Photo by United Air Lines*)

FIG. 118. Weather bureau observers about to determine, by means of a free balloon, the velocity of the upper air currents at different altitudes.

up by the United Air Lines ground stations. Should unfavorable weather develop after the take-off, pilots are advised prior to reaching that zone.

Airway Lighting Beacons. The original standard airway beacon is a 24-inch rotating unit of the search-light type developing approximately 1,000,000 candle-power. The new standard airway beacon is a 36-inch rotating unit which shows two beams

FIG. 119. The pilot of a transport plane reading from the teletype machine a description of the weather conditions along the route over which he is about to take his plane.

of light separated by an angle of 180°, and has a candle-power for each beam of about 1,250,000. The airway beacons are so operated as to show six clear flashes per minute —the single-ended 24-inch beacon revolving six times a minute, and the double-ended 36-inch beacon three times a minute. The 24-inch beacons were spaced at 10-mile intervals. With the more powerful 36-inch lights, however, 15-mile spacing is ample.

Course lights with color screens are used in conjunction with the beacon lights, green flashes being used to indicate the presence of landing facilities and red to show absence of such facilities. Two course lights are mounted on each beacon tower, one pointing forward and the other backward along the airway course.

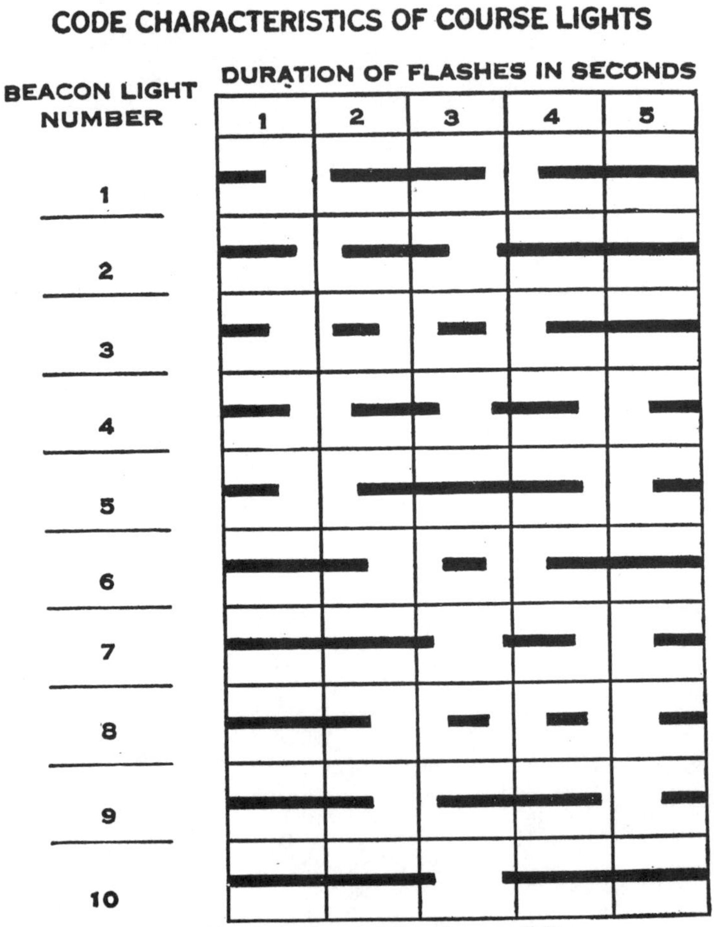

FIG. 120. Code characteristics of course lights.

The course light's code signal (Fig. 120) consists of long and short flashes which identify the beacon by the last digit in its number. For positive identification it is necessary that the passenger know over which 100-mile section of the airway he is flying, inasmuch as the code signals are repeated identically for each ten beacons. For example, the signal for beacons number 1, 21, 31, etc., is the same. These beacons are approximately 100 miles apart.

Where an intermediate landing field is located in close proximity to a hill, it has been found desirable to locate the rotating beacon on this point of vantage, in order

(*Photo by United Air Lines*)

FIG. 121. Talking with Planes in Flight. Operator at a ground transmitting and receiving radiotelephone station, communicating with a pilot in flight. During flight the pilot must report his position every twenty minutes, on regular schedules, indicated by markings on the special clock seen in the picture.

to obtain better long-distance visibility. In this case there generally is located on the field a flashing code beacon light, which flashes the beacon site number according to the code characteristics for course lights. Units of this type sometimes are used to mark special features of the terrain, and in some instances are mounted above rotating beacons instead of course lights. In either case red or green color shades are used in accordance with standard airway practices for indicating the presence or absence of landing facilities.

The electric power required to light the beacons is obtained in several ways. In thickly settled parts of the country electric power is almost everywhere available. In mountains or other locations distant from human habitation, a gasoline-engine-driven generator is used. In isolated regions permanent quarters for airway keepers and their families are provided at intermediate landing field sites. Storage facilities for large quantities of fuel and supplies are necessary to cover winter periods, in some localities six months in duration, when heavy snow makes these sites inaccessible.

From the high points of the Alleghenies, the Rockies, the Wasatch and the Sierra, land-lubber lighthouse keepers prepare each spring to end their winter isolation. These beacon tenders and weather observers generally conclude their hibernation in March, but unusually heavy snows, particularly in the Sierra, keep some of them at their lonely stations until May. Occasionally planes have to drop food to remote stations, where supplies have been exhausted during the long winter siege. These watchers of the mountains and the western deserts are Department of Commerce employees. Those at snowbound mountain posts often have only a waving acquaintance with pilots. The keeper must be weather observer, airport operator, radio mechanic and occasionally host to wayfarers of the air.

The air route from New York to San Francisco, showing the location and number of the airway beacon lights, and also the intermediate and other landing fields, is given on Plate VII.

Airway Radio Stations. There are in all three kinds of radio aids to air navigation, namely (1) radio broadcast stations, for communication by voice; (2) radio range beacons, and (3) radio marker beacons.

The *radio broadcast stations* are ground stations (Fig. 121) which send to and receive from pilots (Fig. 122) of planes in flight reports on weather, dispatching orders, and a check every twenty minutes of the plane's position. In case of necessity, the pilot can report to the ground station more frequently than at the 20 minute intervals. Over this same radio he can also talk to the pilots of other planes in flight. This is short-wave radio. To make possible phone conversation between the plane and the ground, and between pilots of planes in flight there are twenty-two ground radio stations between New York and California. The Department of Commerce also maintains radio communication stations (as shown on Plate IX) at intervals of about 200 miles along the air route, for the distribution of weather reports.

The *radio range beacons* enable the pilot to reach his destination when he is obliged to fly for an extended period out of sight of the ground. The radio range beacons send out code signals which interlock to form the "on course" designation in the narrow bands coinciding with the air route (Fig. 123). The courses are not actually straight lines but narrow V's (Plate XI) that widen out to 8 or 10 miles when 100 miles away from the beacon. The included angle of each course varies from one to four degrees. Radio range beacons are situated about every 200 miles apart, as shown on the maps forming Plates VIII and XI and XII.

The *radio marker* beacons are situated at intervals between the range beacons. These indicate to the pilot in flight as he passes over them, their code designation, thus enabling him from time to time to check up on his position. (See Plate VIII.)

FIG. 122. Pilot talking to ground station 200 miles distant and 12,000 feet below him.

These radio facilities can best be explained by describing what one would receive in making a flight from the Chicago municipal airport to the Cleveland municipal airport (Consult Plates VII and VIII).* "Taking off from the Chicago municipal airport the radio receiver would be tuned to a frequency of 350 kilocycles, where one would hear the Chicago radio range. One would follow the course extending approximately southeast and when on the proper course would hear a series of long dashes (the combination of the code letters A and N) with the Chicago range identifying signal (. . — .) repeated every 12 seconds.

"Should the plane get off the course to the northeast, the long dash would gradually break up and one would hear the off-course signal A (. —). The course should then be corrected and should the plane get off the course to the southwest, the N signal (— .) would be heard. After the plane had been under way about 15 minutes the radio range would stop and the weather broadcasting station would broadcast weather information concerning the various routes emanating from Chicago on the same frequency as the radio range. This will require no change in tuning the receiver, so that the pilot may be always certain of receiving the weather reports as well as any emergency message which it may be necessary to broadcast to him.

After approximately 15 minutes of flying the course would be changed due east and the receiver tuned to a frequency of 320 kilocycles. On this frequency the radio

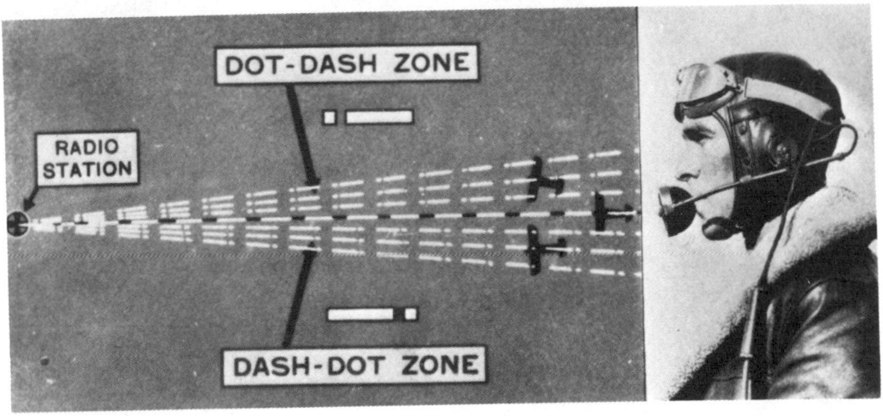

FIG. 123. Radio range beacon. This illustrates the directive radio beam operation. When the plane is to the right of the course the pilot gets a dot-dash; when he is to the left of the true course he gets the dash-dot; when he is on the true course, he hears a steady stream of dashes, similar to the sound heard on a dial telephone, which means he is following his prescribed line of flight, even if he cannot orient himself with ground objects.

range at Goshen would be heard and the pilot can keep on the course by means of the interlock signals from this station. Should the plane drift to the north the interlock signal would break up into the letter A and should the plane drift to the south, the letter N would be heard. The identifying signal on the Goshen range (— — .) will be heard every 12 seconds.

When the plane is within about 5 miles of McCool, the marker beacon will be heard sending a signal of two dots at 5-second intervals. This signal will increase in intensity until the station has been passed and will then gradually decrease until the plane is out of range.

(The distance over which the marker beacon stations can be heard varies quite materially with the seasons of the year. Effort is made to keep them adjusted so that

* From U. S. Dept. of Commerce, Aeronautics Branch, General Airway Information Bulletin No. 1, 1932.

.heir range will not exceed 3 miles, but sudden changes in conditions affecting radio transmission often occur which in many cases will double or triple the range.)

The plane would continue on its course until it arrives over the Goshen station, the signal strength increasing as the station is approached until it will probably be necessary for the pilot to reduce the volume of the signal at frequent intervals so that it will not be uncomfortable to the ears.

After the plane passes over the Goshen station, a reversal takes place in the course signals. While to the west of Goshen the A signal (.—) was to the north of the course, after passing over Goshen the A signal is found to the south of the course, this being an inherent characteristic of the radio range and is not subject to change. As the plane passes immediately over the radio range antenna system, there will be a complete fade-out of the range signals. This is caused by a silent zone directly over the range antenna system. However, since this station is not operated continuously but is operated on a one minute on, one minute off-schedule, it may not be possible to note the silent zone because the range may have been shut down during the time the plane passed over it.

As the plane proceeds on its course, it would pick up the marker beacon at Helmer (...). During the time that the radio receiving set has been tuned to the Goshen radio range, weather reports will be received at periodic intervals from the radio station at Jackson, Mich., which will give information concerning the weather conditions along the airway between Chicago and Cleveland. A short time after passing the Helmer marker beacon, the receiving set should be tuned to a frequency of 344 kilocycles, on which frequency will be heard the Cleveland radio range and the Cleveland weather broadcasting station. Since the radio courses emanate from the station in a straight line, a portion of the course is over the lower end of Lake Erie. A marker beacon has been placed at Vickery to indicate the presence of a stretch of water.

Flying by Night. To the uninitiated, night flying probably holds the greatest charm of air travel.

As the pilot takes off, amid the lights that flood the field, for a Division point three to four hundred miles away, a code message goes out over the teletype circuit reading:

"United Air Lines plane license No. NC413H, Pilot Lewis, departed Newark twelve midnight for Cleveland."

This message is automatically reproduced on receiving machines at strategic points on the airways to check the plane's progress.

Soon after leaving the ground the pilot is greeted by a flash of clear light from the next beacon beyond the field that he is leaving. As it revolves, a red light, known as a course light, flashes a dot-dash code signal. As the red code signals discontinue the revolving beacon's white flash again appears. This continues from sundown to sunup on the chain of beacons spaced about ten miles apart.

Every third beacon light has a green course light which tells the pilot that at this beacon there is a Department of Commerce intermediate landing field, marked by boundary lights 250 feet apart. These lighted emergency fields are maintained so that planes can make landings between terminals, if advisable or necessary.

But the pilot may rely on more aids in night air navigation, including the directive radio beacon, already described. Through ear phones he listens to the signals from radio beacon beams to keep him on his course at all times. If the pilot is a little off his course to one side the dot-dash signals predominate in strength, or if to the other side the dash-dot signals come in more distinctly. If he is exactly on his course the signals merge into one long dash.

Therefore, without watching beacon lights, or if poor visibility obscures them, he may follow a true course by flying his plane so that the long dash predominates in his ear phones. Every group of signals is followed by the identification signals of the station to which he has been listening.

When there is no occasion to land, the pilot continues on his route watching the beacon lights, listening to the radio range beacon and checking his flight and engine instruments. He is also listening to the radio telephone and occasionally reporting his position to ground stations. Occasionally the radio beacon signals cease and are followed by a voice announcing the station, the time, the ceiling at the field at which the plane will next land, the condition of the weather, wind velocities, temperature, barometer reading and all other information of value and assistance to a pilot flying in the night or day.

To the passenger night travel has a fascination not felt by those who travel only by daylight. In mountainous country the dim outlines of the near-by masses rise up bold and black beside the course and seem far greater than under the glare of the sun. With the gradual waning of the evening light the exhaust from the motors alongside the cabin spits out sheets of flame which throw a lurid glow upon the wings and struts just outside of the window.

The flashing beacons off at one side are then easily seen, and to the night traveller over the desert they constitute the only features of interest outside of the cabin. A half dozen of them may easily be visible at a time, lighting the route 40 or 50 miles behind and 20 or 30 miles ahead of the plane. To the pilot still more are to be seen because of his more advantageous position. Towns occasionally glide by, far beneath, but also visible at great distances. Trains with rows of sparkling lights, long transcontinental limiteds crossing the western plains, may show where the route coincides with a main railroad line. Automobile headlights on the highways mark out the roads almost as clearly by night as by day.

At night, too, there is more than ordinary interest in making a landing. There is no doubt now as to the location of the field, with its border of twinkling lights and its center of darkness. As usual, before landing the pilot takes the plane directly over the hangar to note from the large illuminated arrow the direction of the wind near the ground. Prior to reaching the field he has also been advised over the radio telephone of the wind conditions and this information is repeated to him by the ground radio station as he circles for a landing. Sweeping majestically two or three miles into the outlying country, the plane returns and settles down onto a ribbon of white, sprayed there by the landing light under the pilot's control.

VI. PREPARATIONS FOR AN AIR TRIP

Maps. Preparations for a trip by air should be very much the same as for any kind of journey if the greatest profit is to accrue from the trip. The first consideration is that of maps. Unquestionably, the air traveller should be equipped with a full set of all the airway maps of the route if he wishes more detailed map material than is afforded by the airlogs distributed by the air lines to all passengers. Strip airway maps (Plate X), issued by the U. S. Coast and Geodetic Survey, cover all the important airway routes of the country. These, however, are being replaced by "Sectional Maps" which, instead of covering definite air lines, cover definite sections of the country. Eighty-seven of them will eventually encompass the entire United States. The traveller planning to go by air will therefore do well to communicate with the U. S. Coast and Geodetic Survey, Washington, D. C. (or 6 State St., New York City) and ask for an index map showing what sectional and strip airway maps have been published. The maps covering the route should then be purchased and folded in accordion style so as to facilitate easy consultation.

The small scale of the airway maps, namely 1 to 500,000, or about eight miles to an inch, makes it impossible to show the details of topography and culture which the student of geology is accustomed to find on topographic maps. The generalization in the matter of relief will be a cause of annoyance, especially in the eastern United States, where the beautiful details of the folded Appalachian ridges are all but lost on the map. In other words, the airway maps cannot be interpreted in terms of structure as can be done with the topographic maps of the U. S. Geological Survey. Another disconcerting matter is the fact that only the main roads are shown. This is especially troublesome in parts of the country like northern Ohio, Indiana, and Illinois, where roads follow every section line and are only a mile apart. One sees numerous roads on the landscape which are not shown on the map but which look fully as important and conspicuous as those that are depicted. The identification of towns, too, is at first difficult. Most of the smaller towns and villages are represented merely by circles on the map. Their characteristic shapes, which might help in distinguishing them, are not given, but more and more cities are coming to the aid of the air traveller by painting the name of the city on some strategic building. It is no wonder, therefore, that even those who are accustomed to the use of maps frequently fail to identify many features when they first fly over them. However, these difficulties disappear with practice, and the air traveller soon becomes as facile as the pilots in using the airway maps. The features which are most helpful to pilots are large rivers, railroads, important highways, and prominent topographic forms. The pilot taking over a new route, with which he is not familiar, first rides as a passenger several times until he is acquainted with the landmarks. Then he makes three trips alone as a mail pilot over the route before he is entrusted with passengers. Pilots of transport planes do not know the names of all the towns even along their regular runs, but they do know the precise location of the plane during flight. Therefore, realizing the difficulty of others, the person riding over a route for the first time need not become discouraged over difficulties he himself may have in keeping his location at every step of the journey or because he makes mistakes in identifying places. On his second trip many of these difficulties disappear.

The air traveller will find it good practice as he rides over an established route to look for the various landing fields along the way. These are indicated by appropriate symbols on the airway maps and are marked on the ground by yellow circles and other devices. The Department of Commerce intermediate field station buildings have a number painted on the roof and frequently these identify the location of the air traveller. Even with a knowledge of almost the precise spot where a landing field is located, the inexperienced traveller will have difficulty in detecting it from the air. Pilots have this same difficulty when coming to a field for the first time. Perhaps the chief reason for the difficulty which the inexperienced air traveller has in recognizing landing fields is their small size, as viewed from aloft.

All the airway maps, both strip and sectional, show clearly the established lighted airways, with the location of all emergency landing fields and beacons. The lighted route is almost everywhere accessible to railroad or road, so as to facilitate the upkeep of beacons and fields. The flashing beacons are located about every ten miles, the landing fields about every thirty to fifty miles. Constant additions are being made to the number of both the fields and the beacons, and these facts are recorded upon the latest editions of the airway maps. In rugged mountain passes, as across the Sierra Nevada, and in other places where the terrain is very uneven, beacons may be spaced every mile or two. These are all shown on the map and are readily noted by the air traveller especially at night.

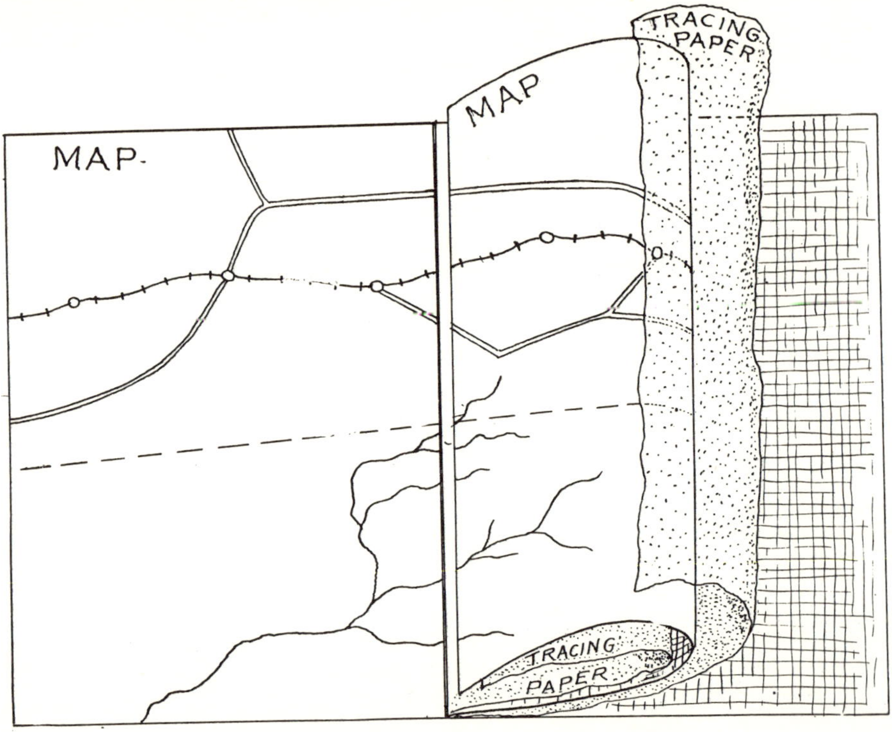

FIG. 124. Sketch showing arrangement of maps in notebook.

For the geologist the state geological maps are very convenient in scale and will be found a source of satisfaction, notably those for New Jersey (Dept. of Conservation, Trenton), Pennsylvania (Bureau of Publications, Harrisburg), and Wyoming (U. S. Geological Survey, Washington). The U. S. Geological Survey topographic sheets are discussed in the next chapter. For general use the maps of the different states published by Rand McNally and Company for use in air travel may be recommended. The ordinary automobile road maps distributed by the oil companies are simple to read and usually the roads are easy to distinguish from the air. People already familiar with this type of map would probably find them more satisfactory than any other kind for following the course of the plane across the country.

Notes and Photographs. The note-book, if any, depends largely upon the peculiar interests and bent of the traveller. In any event it should be conveniently and invitingly arranged so as to encourage rather than repel note taking. There are difficulties in making notes and memoranda, difficulties inherent in the task itself, as well as the more subjective ones due to the natural lethargy of the observer. An incentive to activity is sometimes necessary.

The note-book illustrated in Fig. 124 has been found satisfactory in almost every respect. It is a standard $7\frac{1}{2} \times 10$ inch stiff-covered book with cross-section ruling. The airway maps are cut into pieces of this size and pasted on the left hand pages of the notebook. Only the inner margin of the map need be pasted, so that the map really forms another leaf in the book. Then over each map is placed a piece of tracing paper and this is pasted also only along the center line of the book. The book thus consists of map, tracing paper, cross-section page, map, tracing paper, cross-section page, etc., in that order. The maps are arranged with the western ones in the front of the book, the eastern ones coming in succession, so as to permit the turning of pages as shown in the sketch in order to bring the corresponding edges of the different maps together when passing from one map to the next.

Notes are taken on the maps and on the note-book page facing the map. The tracing paper is useful for marking the exact location where photographs are taken and the direction of the view, as well as for indicating the position of other things observed. A full note-book page is ample for the notes which pertain to the map which it faces. At the end of each trip the note-book pages covered over by the maps are used for a more leisurely exposition of things seen but not sufficiently recorded during the actual flight.

The taking of photographs with almost any kind of a camera is quite practicable. A ray filter is strongly recommended but it should not be too dense. It is, however, necessary to realize that pictures taken from the air need very short exposures. The danger of overexposure is far greater than underexposure. The speed of the plane hardly has to be considered except for pictures taken very close to the ground. Probably the best rule to follow is to adopt about the same combination of filter, time, and diaphragm opening, as would be suitable when taking distant views of a bright landscape or open body of water.

Just before the flight starts it is well to have the note-book in readiness, all extra maps (geological and such) close at hand and arranged in the right order for quick reference, extra camera films unwrapped and ready for immediate insertion, and two or three pencils nicely sharpened. Never mind, however, what the other passengers may think about these preparations.

And, finally, do not feel that you have wasted your time in making these memoranda if you do not use them later. They have served their purpose when you have made them, and you will in the future review your journey largely in terms of those particular things which you have recorded. But at the same time, do not take it all too seriously. There is beauty in the form of clouds, there is ecstacy in riding mile-high above a glorious landscape, and there are many things which can not be caught with a pencil or camera, but which you will treasure ever after as you look back upon a rare and interesting experience.

VII. MAPS

Selected List and Description of Topographic Maps Traversed by the Air Route

Plate VI shows the location and names of the topographic sheets published by the United States Geological Survey, that are traversed by the air route. It is impracticable and perhaps undesirable for most travellers to carry many of these maps with them during flight. The following selected list is designed to draw attention to those of particular interest. The person familiar with topographic maps will derive considerable satisfaction in having some maps before him during the journey. About three dozen have been selected for that purpose. When it is remembered that a topographic sheet, on the usual scale of 1: 62,500 is traversed in 6 or 7 minutes by most planes it becomes obvious that not much time is permitted to make observations. Nevertheless, during this brief time it is possible to recognize an astounding lot of detail and to gain an impression of the country not easily eradicated. The descriptions which follow are arranged in proper sequence from the Atlantic to the Pacific coast.

In New Jersey the several U. S. G. S. topographic sheets indicated on the Index Map (Plate VI) are no longer to be secured from the U. S. Geological Survey, but may usually be consulted in libraries. The area, however, is covered by similar maps issued by the New Jersey State Department of Conservation and Development. The maps sell at 50 cents apiece, each sheet covering about the equivalent of four U. S. Geological Survey topographic sheets. Three of the large maps cover the New Jersey portion of the air route.

New Jersey Map No. 26 includes approximately the area covered by the Harlem, Brooklyn, Paterson, and Staten Island sheets of the U. S. Geological Survey; No. 25 covers the Morristown, Plainfield, Lake Hopatcong, and Somerville sheets; and No. 24 covers most of Hackettstown, High Bridge, Delaware Water Gap, and Easton sheets.

New Jersey Sheet No. 26 shows the highly developed region around New York City. To the traveller starting from or arriving at Newark by air it is a very useful map to have in hand. All the ramifications of New York Harbor come within the scope of this map. The long ridge of the Palisades and the two Watchung ridges, none of which is specifically named on the map, are the most prominent topographic features of the area. The steep eastern slope of the Watchungs and the gap at Milburn with its numerous morainal hills are clearly shown on the map and easily identified from the air.

New Jersey Sheet No. 25 shows the southern part of the Watchung Mountains curving toward the south. Long Hill, a third ridge, also appears. The air route passes between Long Hill and Second Watchung Mountain. The outer steep slopes are well shown. Numerous trap quarries, visible from the air, are indicated. The western part of the area shows the irregular topography of the crystalline Highlands. South of Plainfield and Bound Brook are the flat plains of outwash, spread westward from the terminal moraine which is excellently represented in the southeastern corner of the map. Hadley Field near Bound Brook is situated on this plain.

New Jersey Sheet No. 24 shows the system of ridges making up the Highlands. Cushtunk Mountain, an arcuate trap ridge, with its forest cover, appears near the eastern margin of the area. The air route passes directly over the small towns of Lebanon and Annandale and along the northern side of Cushtunk Mountain. From the air the towns of High Bridge, Junction, and Washington may easily be identified. The broad valleys of Musconetcong and Pohatcong Creeks are developed on infolded belts of limestone.

The **Allentown, Penna.,** sheet is traversed in its northern part by the air route which goes almost over the town of Bath. This map shows in its southeastern part the crystalline hills of the Reading Prong, in its middle part the open flat-floored limestone portion of the Great Valley, and in its northern part the rolling shale or slate belt which makes up the western half of the Great Valley.

The **Mauch Chunk, Penna.,** sheet. The route crosses the southern part of this map from Slatington to Ashfield. Blue Mountain is a continuation of Kittatinny Mountain. The very resistant Shawangunk conglomerate making up this ridge is succeeded westward by other overlying beds of lesser resistance, mostly Devonian in age. To the north the heavily forested Pocono Plateau is underlain by the horizontal beds of Pocono sandstone (Mississippian).

The **Hazelton, Penna.,** sheet illustrates a particularly interesting part of the anthracite coal field. The route crosses the southern end of this map almost over the towns of Coaldale and Tamaqua. Mauch Chunk Ridge consists of Pocono sandstone dipping northwest. Pisgah Mountain and Nesquehoning Mountain are the limbs of a syncline pitching southwest, formed by the resistant Pottsville conglomerate, and enclosing along its axis the Coal Measures. Broad Mountain in the center of the map is an anticline of Pocono sandstone pitching southwest.

The **Mahoning, Penna.,** sheet covers an unusually complicated part of the anthracite coal field. The southern part of the map is essentially a broad syncline with a number of anticlinal axes. Along the synclinal axes the coal beds occur. This accounts for the position of Shenandoah, a city directly under the air route.

The **Catawissa, Penna.,** sheet. The air route passes through the center of this map over the town of Numidia. To the north is the nose of an eastward pitching syncline, its limbs represented by Nescopeck Mountain and Catawissa Mountain which in turn swings around to the west again to form Little Mountain, where the rocks dip southward. All of these are formed by the resistant Pocono sandstone. Numidia lies in a broad basin of Devonian shales developed on the axis of an anticline. The southern part of the map is mainly a syncline with the city of Mount Carmel on its axis where the Coal Measures occur. Big Mountain and Mahanoy Mountain of Pottsville conglomerate are the two limbs of this structure.

The **Shamokin, Penna.,** sheet shows the western end of the syncline just mentioned, with Shamokin occupying its axis. The air route runs east and west through the center of this map over the valley of Shamokin Creek.

The **Centre Hall, Penna.,** sheet shows two of the large anticlinal limestone valleys which characterize the western part of the folded belt, namely Nittany Valley and Penn Valley. Between the two valleys lies Nittany Mountain, the two crests of which constitute the limbs of a northeastward pitching syncline. The limestone character of the valleys is suggested by the depression contours. The air route passes over the center of the map just north of Penn Cave and provides a good view of Brush Mountain which is the nose of another syncline pitching toward the northeast. All of the major ridges on this map are formed by the resistant Bald Eagle conglomerate (Silurian) which overlies the Ordovician limestones.

The **Bellefonte, Penna.,** sheet is traversed in its northern part by the air route. The town of Bellefonte, as well as the Bellefonte airport east of the city, lie in the Nittany Valley, developed on a broad anticline. Nittany Mountain is the nose of a northeastward pitching syncline. Bald Eagle Ridge is formed of resistant Silurian conglomerate dipping northwestward beneath the plateau series. In front of the plateau proper is a dissected bench underlain by Devonian shales, the plateau itself rising in a bold escarpment, formed mainly by the Pocono sandstone.

The **Clarion, Penna.,** sheet shows a very typical part of the non-glaciated portion of the Allegheny plateau. The Clarion River and its tributaries are deeply incised below the plateau surface. The air route crosses the northern part of this map directly over the city of Clarion. The Foxburg-Clarion Folio, No. 178, describes this area.

The **Franklin, Penna.,** sheet is traversed by the air route just north of its southern margin. The incised meanders of the Allegheny River are the most striking features of the region. The contrast between the undercut and slip-off slopes of the meander spurs are represented here better than usual on topographic maps. The Kennerdell intermediate landing field is situated on top of the broad meander spur just east of Kennerdell.

The **Stoneboro, Penna.,** sheet lies in the glaciated part of the plateau and offers a striking contrast with the maps lying to the east. The terminal moraine, with its occasional ponds, runs north and south covering practically the eastern half of the map and is especially obvious southeast of Stoneboro. The air route crosses this part of the moraine as well as the large marsh lying just to the west of it.

The **Berea, Ohio,** map, upon which the Cleveland airport is situated, is especially interesting because it shows the location of several of the beaches of former Lake Maumee. North Ridge, Middle Ridge, Chestnut Ridge, and Butternut Ridge are all distinctly shown and named. On the crest of each one is an important highway. The air route crosses the map practically over the junction of East and West Branches of Rocky River, at which point may be seen two interesting interconfluent hills formed by successive changes in the point of junction.

The **Sandusky, Ohio,** sheet shows part of the bed of Glacial Lake Maumee. The air route crosses the middle of the map over Huron and Bogart. South of the route and just too far away to be easily seen from the air are patches of sand dunes covering several square miles of country. Sandusky Bay with its long protecting spit is another feature of interest.

The **Pioneer and Bryan, Ohio,** maps together show remarkably well the characteristics of the till plain of northern Ohio, although the southeastern half of the Bryan sheet covers the basin of Glacial Lake Maumee. Between Bryan and Hicksville several beach ridges along the old lake shore may be noted. The air route passes almost over the city of Bryan. The Pioneer sheet shows part of the Wabash moraine crossing its northwestern corner. This illustrates nicely the characteristic topography of the moraines to be seen in northern Indiana for which detailed maps at the present are not available. Across the middle of the Pioneer map runs the "Fulton Line" along which many of the north-south roads are offset, this being one of the correction lines in the land office survey. The checkerboard plan of the road system in this part of the country is noteworthy.

The **Englewood and Berwyn, Illinois,** sheets on the scale of 1: 24,000 will be found especially interesting to the traveller as he arrives at and leaves the Chicago airport. The airport covers part of both maps, being just within the Chicago city line, south of the Laramie Golf Club, at the corner of West 63rd Street and South Cicero Avenue. The most striking topographic feature on the two maps is the Chicago Sanitary and Ship Canal which runs through the heart of the city. The beach ridges of former Lake Chicago, crossing the airport, are clearly shown by the contours.

The **Dixon, Illinois,** map shows a typical part of the prairie plains covered by the old Illinoian drift. The country is extremely flat except near the major streams where dissection has produced mature topography. The air route crosses the southern part of the map.

The **Des Moines, Iowa,** sheet will prove of interest mainly because it shows the environs of the largest city in Iowa seen along the air route. The broad flood plains of Des Moines and Raccoon Rivers are the most prominent topographic features. The airport is several miles northeast of Des Moines, just south of the town of Berwick. The former course of Des Moines River east of the city is clearly shown by the wide flood plain at Highland Park and Easton Place.

The **Omaha and vicinity, Nebraska, and Iowa,** sheet shows the broad flood plain of the Missouri River. The Omaha airport is situated northeast of the city between Florence Lake and Cutoff Lake. The flood plain of the river is 4 to 5 miles in width and includes several other cut-off lakes. The steep loess-covered bluffs on the east side of the river are clearly shown.

The **North Platte, Nebraska,** sheet shows the junction of the North Platte and South Platte Rivers. The North Platte airport is siuated east of the North Platte River and very close to it, south of the railroad and highway. The northern part of this map illustrates a very characteristic part of the Sand Hill region of Nebraska with its numerous irregular depressions and small elevations. Of especial interest on this map is the contrast between the single broad channel of the North Platte and the braided pattern of the South Platte.

The **Cheyenne, Wyoming,** map reveals by the intricacy of its contours the minutely irregular surface of this part of the Great Plains. There is a fairly uniform slope from an elevation of 6,600 feet in the western part of the region to 5,700 in the eastern, a slope of about 900 feet in 25 miles or 36 feet to the mile. The surface, however, is cut by numerous dry stream courses. The airport is situated close to the northern edge of the city.

The **Rock Springs, Wyoming,** sheet includes only a small part of the Baxter Basin. The axis of the Rock Springs uplift runs through the eastern part of the map, where erosion has reduced the region to a lowland. Running north and south through the western part of the map is one of the encircling ridges, or rather belts of hills, due to the Mesa Verde formation, dipping here toward the west. The Rock Springs airport is situated practically on the western edge of the map two or three miles north of the city. Of especial interest in this area is the fine-textured badland topography so characteristic of this arid region.

The **Salt Lake, Utah,** sheet is not especially satisfactory because of its small scale (1: 250,000) and the fact that it is one of the early reconnaissance maps first published in 1885. Nevertheless it gives a fair conception of the form of the Wasatch Range with its steep western side. The air route goes over Coalville, then through one of the passes and down a canyon to the airport west of the city.

The **Carson Sink, Nevada,** sheet. The air route crosses the northwestern corner of this map, skirting the south side of Humboldt Range. In spite of its small scale (1: 250,000) and its large contour interval (100 feet) the map shows clearly the remarkable bar at the west end of Humboldt Lake. Carson Sink, represented as a large body of water, may at times be completely dry due to the use of all the water for irrigation.

The **Truckee, Calif.,** sheet depicts the main crest of the Sierra Nevada where it is crossed by the air route. Although the lighted airway follows the Yuba River valley, in clear weather the planes fly several miles to the south practically over the North Fork of American River. The effect of glaciation is suggested by the cirques along the crest of the range and the numerous lakes.

The **Colfax, Calif.,** sheet covers part of the back slope of the Sierra upland. The air route crosses the map just south of North Fork of American River whose canyon is one of the most striking features on this map. The remarkable flat divides between the canyons are remnants of the Tertiary lava flows which buried the previous irregular topography.

The **Folsom, Calif.,** map (scale 1: 31,680) with a contour interval of 5 feet is an example of the placer mining region along the American River. The location of gravel dumps is clearly shown.

The **Honker Bay,** the **Collinsville,** as well as the **Jersey, Calif.,** sheet, all on the large scale of 1: 31,680 show in great detail part of the San Joaquin River valley, much of which has been reclaimed by diking. Pumping plants along the river lift the water from the polder lands over the protecting dike into the river. North of the river, and well shown on the Collinsville sheet is a rough bit of topography constituting the Montezuma Hills, over which the air route passes. This is a region of older alluvium, uplifted and now maturely dissected.

The **Haywards, Calif.,** sheet shows some of the alluvial flats and marshlands adjacent to San Francisco Bay. The Oakland airports are situated along this shore.

VIII. REFERENCES

Suggestions for Reading

The following titles constitute, in the opinion of the author, the most suitable references for the general reader. Most of these may profitably be examined before undertaking an air tour across the country. There are scores, indeed hundreds, of other books, papers, and articles in scientific journals, besides those given below. The student of geologic literature will have no especial difficulty in finding them in any well-equipped geological library.

Many of the books noted below contain bibliographies which will introduce the student still further into the literature of the subject.

I. General references on Aeronautics.

1. U. S. Dept. of Commerce, Aeronautics Branch. General Airway Information. Airway Bulletin No. 1. 1931, revised 1932.
   An extremely useful handbook containing maps showing light and radio beacons for all the air routes of the country.
2. U. S. Dept. of Commerce, Aeronautics Branch. Descriptions of Airports and Landing Fields in the United States. Airway Bulletin No. 2, 1931, revised 1932.
3. Smith, Wesley L. Air Transport Operation. New York, McGraw-Hill Book Co., Inc., 1931.
4. Maguire, Charles J. Aerology. 136 pp., New York, McGraw-Hill Book Co., Inc., 1931.
   A non-mathematical discussion of the atmosphere, intended to assist aviators in understanding weather phenomena.
5. Holland, Harvey H. Avigation. 270 pp., New York, McGraw-Hill Book Co., Inc., 1931.
   A simple text-book for pilots, but useful as a general reference for any one interested in the subject of air navigation.
6. Weems, P. V. H. Air Navigation. McGraw-Hill Book Co., Inc., 1931.
   A comprehensive book of reference, containing a long and useful bibliography.
7. Matthes, Gerard H. Aerial photography as an aid in geological studies. Am. Inst. Min. and Met. Eng., Trans. Vol. 76, 321–336, 1928.
8. Reeves, D. M. Aerial Photographs. New York, Ronald Press, 1930.
   Explains how to recognize ground features on an aerial photograph. Many illustrations.
9. Lee, Willis T. The face of the earth as seen from the air. Am. Geog. Soc., Special Pub. No. 4, 110 pp., 1922.
10. Miller, O. M. An experimental air navigation map. Geographical Review. Vol. XXIII, pp. 48–60, 1933.
    The map covers the Pittsburgh-Cleveland area.
11. Curry, Manfred (Ed.). The Beauty of Flight. 12 pp. 100 pl., New York, Day, 1932.

II. The United States as a Whole.

1. Bowman, Isaiah. Forest Physiography. 750 pp., map, New York, John Wiley & Sons, 1911.
   A text-book describing the physical features of the entire United States.
2. Lobeck, A. K. Physiographic diagram of the United States. Scale, 1: 3,000,000. Chicago, A. J. Nystrom & Co., 1921.
   Shows, as in a bird's-eye view, the physical features of the country, with their names.
3. Fenneman, N. M. Physiographic Divisions of the United States. Assoc. Am. Geographers, Annals, Vol. 18, No. 4, pp. 261–353, map, 1928.
   The map is especially valuable as it portrays the physical divisions of the country with accompanying brief descriptions.
4. Davis, W. M. Guidebook for the Transcontinental Excursion of 1912. American Geographical Society of New York. 144 pp., 1912.
5. Fenneman, N. M. Physiography of the Western United States. 534 pp., 173 figs., New York, McGraw-Hill Book Co., Inc., 1931.
   A standard text-book describing the region west of the Mississippi River.
6. Lee, W. T., and others. Guidebook of the Western United States, Part B, The Overland Route. U. S. Geol. Survey, Bull. 612, 1915.
   Describes the route of the Union Pacific Railroad from Omaha through Cheyenne and Ogden to San Francisco.
7. Ward, Robert DeC. The Climates of the United States. 518 pp., New York, Ginn & Co., 1925.
   Describes weather, climate; effect on health, crops, and other geographical aspects.
8. Kincer, J. B. Precipitation and Humidity (in the United States). Atlas of American Agriculture, Part II, Sect. A, U. S. Dept. of Agriculture, 1922.
   Numerous maps and text. The most convenient source of information upon rainfall distribution in the United States.

9. Kincer, J. B. Temperature, Sunshine, and Wind (in the United States). Atlas of American Agriculture, Part II, Sect. B, U. S. Dept. of Agriculture, 1928.
   Many colored maps and text.
10. Shantz, H. L., and Zon, Raphael. Natural Vegetation (of the United States). Atlas of American Agriculture, Part I, Sect. E, U. S. Dept. of Agriculture, 1924.
   Splendidly illustrated with colored map and many photographs. Abundant text.
11. Baker, O. E. A Graphic Summary of American Agriculture. U. S. Dept. of Agriculture, Misc. Pub. No. 105, 1931.

III. **New Jersey.**

1. Merrill, F. J. H. and others. The New York City Folio. U. S. Geol. Survey Folio 83, 1902.
2. Bayley, W. S., and others. The Raritan Folio. U. S. Geol. Survey Folio 191, 1914.
3. Davis, W. M. The Rivers of Northern New Jersey. Nat. Geog. Magazine. 2: 81–110, 1890. Also in Geographical Essays, Boston, 1909.
   Gives a clear and valuable interpretation of the scenery of northern New Jersey, especially the Watchungs and the Highlands.
4. Johnson, Douglas. Stream Sculpture on the Atlantic Slope. 142 pp., New York, Columbia University Press, 1931.
   Presents the most recent explanation of the topography of Northern New Jersey.

IV. **Pennsylvania.**

1. Davis, W. M. The Rivers and Valleys of Pennsylvania. Nat. Geog. Mag. 1: 183–253, 1889. Also in Geographical Essays, Boston, 1909.
   A technical discussion of the origin of the relief features in the folded Appalachians of Pennsylvania.
2. Willis, Bailey. The Northern Appalachians. Nat. Geog. Soc. Monograph 1, No. 6, 169–202, 1895. Also in The Physiography of the United States. (Nat. Geog. Soc.) 169–202, New York, American Book Co., 1896.
3. Johnson, G. F. Agriculture in Pennsylvania, a study of trends, county and state, since 1840. Pennsylvania Dept. of Agriculture, Bull. Vol. 12, No. 15. General Bulletin No. 484, 1929.
4. Shaw, E. W., and others. The Foxburg-Clarion Folio. U. S. Geol. Survey Folio 178, 1911.
5. Wright, G. F. The Glacial Boundary in Western Pennsylvania, Ohio, Kentucky, Indiana, and Illinois. U. S. Geol. Survey, Bull. 58, 1890.

V. **Ohio.**

1. Leverett, Frank. Outline of the History of the Great Lakes. Michigan Acad. Sci. Rept. 12: 19–42, 1910.
2. Leverett, Frank. Glacial Formations and Drainage Features of the Erie and Ohio Basins. U. S. Geol. Survey Monograph 41, 1902.
3. Carney, Frank. The Raised Beaches of the Berea, Cleveland, and Euclid Sheets, Ohio. Denison University, Sci. Lab. Bull. 14, 262–287, 1909.

VI. **Indiana and Illinois.**

1. Malott, Clyde A. The Physiography of Indiana. In Handbook of Indiana Geology (Indiana Dept. Conservation, Pub. No. 21) pp. 59–256, Indianapolis, 1922.
2. Barrett, Edward. The Dunes of Northwestern Indiana. Indiana Dept. Geol. and Nat. Res. 41st Ann. Rept., 11–27, 1917.
3. Dryer, Charles R. Studies in Indiana Geography. First series, rev. ed., 114 pp. Terre Haute, Ind., 1907.
4. Tucker, W. M. History of the Lakes near LaPorte, Indiana. Indiana Acad. Sci., Proc. 38th Ann. Meeting, 1922, pp. 83–94, 1923.
5. Leverett, Frank, and Taylor, F. B. The Pleistocene of Indiana and Michigan and the History of the Great Lakes. U. S. Geol. Survey Monograph 53, 1915.
6. Young, E. C., and Elliott, F. F. Types of Farming in Indiana. Purdue University, Agric. Exp. Station, Bull. 342, 1930.
7. Leverett, Frank. The Illinois Glacial Lobe. U. S. Geol. Survey Monograph 38, 1899.
8. Alden, W. C. The Chicago Folio. U. S. Geol. Survey Folio 81, 1902.
9. Salisbury, Rollin D., and Alden, W. C. Geography of Chicago and its Environs. Geog. Soc. Chicago, Bull. No. 1, 63 pp., 1920.
10. Fryxell, F. M. Physiography of the Region of Chicago. Univ. of Chicago Press, 55 pp., 1927.
11. Woodard, John. Origin of Prairies in Illinois. Ill. State Acad. Sci., Trans. Vol. 16, 259–263, 1923.

VII. **Iowa.**

1. Holmes, C. L. Types of Farming in Iowa. Iowa State Coll. of Agric., Agric. Exp. Station, Bull. 256, 1929.
2. Schoewe, W. H. Origin and History of Extinct Lake Calvin. Iowa Geol. Survey, Bull. 29, 49–222, 1924.

3. Shimek, B. The Loess of Iowa City and Vicinity. Iowa Univ. Lab. Nat. Hist., Bull. 5, 195–212, 1901. Am. Geol. 28, 344–358, 1901.
4. Lees, J. H. The Des Moines Valley. Iowa Geol. Survey, Bull. 25, 423–615, 1916.
5. Shimek, B. The Prairies. Iowa Univ. Lab. Nat. Hist., Bull. 6, 169–240, 1911.
6. Kay, George F. Numerous papers upon the glacial features of Iowa.

VIII. **Nebraska.**

1. Johnson, W. D. The High Plains and Their Utilization. U. S. Geol. Survey, Ann. Rept. 21, Part 4, 1900, Ann. Rept. 22, Part 4, 1901.
2. Darton, N. H. Preliminary Report on the Geology and Water Resources of Nebraska West of the One Hundred and Third Meridian. U. S. Geol. Survey, Prof. Pap. 17, 1903.
3. Darton, N. H. Preliminary Report on the Geology and Underground Water Resources of the Central Great Plains. U. S. Geol. Survey, Prof. Pap. 32, 1905.
4. Darton, N. H. The Structure of Parts of the Central Great Plains. U. S. Geol. Survey, Bull. 691a, 1919.
5. Hedges, Harold, and Elliott, F. F. Types of Farming in Nebraska. Univ. of Neb., Coll. of Agric. Exp. Station, Bull. 244, 1930.
6. Condra, G. E. Development of the Platte River Bottomland in South Central Nebraska. Assoc. Am. Geographers, Annals, Vol. 21, No. 2, pp. 101–105, 1931.

IX. **Wyoming.**

1. Davis, W. M. The Colorado Front Range. Assoc. Am. Geographers, Annals, 1, 21–84, 1911.
2. Darton, N. H., and others. The Laramie-Sherman Folio. U. S. Geol. Survey Folio 173, 1910.
3. Darton, N. H., and Siebenthal, C. E. Geology and Mineral Resources of the Laramie Basin, Wyoming. U. S. Geol. Survey, Bull. 364, 1909.
4. Schultz, A. R. Oil Possibilities in and around Baxter Basin, in the Rock Springs Uplift. U. S. Geol. Survey, Bull. 702, 1920.
5. Sears, J. D. Geology of the Baxter Basin Gas Field. U. S. Geol. Survey, Bull. 781, 13–27, 1926.
6. Kemp, J. F., and Knight, W. C. Leucite Hills of Wyoming. Geol. Soc. Am. Bull. 14, 305–336, 1903.

X. **Utah.**

1. Lee, Willis T. Geography, Geology and Physiography of the Great Salt Lake Basin. U. S. Geol. Survey, Water-Supply Paper 517, 1924.
2. Davis, W. M. The Wasatch, Canyon, and House Ranges, Utah. Harvard Coll. Mus. Comp. Zool., Bull. 49, 17–56, 1905.
3. Gilbert, G. K. Lake Bonneville. U. S. Geol. Survey, Monograph No. 1, 1890.
4. Atwood, W. W. Glaciation of the Uinta and Wasatch Mountains. U. S. Geol. Survey, Prof. Pap. 61, 1909.
5. Schneider, Hyrum. A Discussion of Certain Geologic Features of the Wasatch Mountains. Jour. Geol. Vol. 33, 28–48, 1925.
6. Hintze, F. F., Jr. Geology of the Wasatch Mountains. N. Y. Acad. Sci. Ann. 23, 85–143, 1913.

XI. **Nevada.**

1. Davis, W. M. The Mountain Ranges of the Great Basin. Harvard Coll. Mus. Comp. Zool., Bull. 42, 129–177, 1903.
A classic treatment of this region.
2. Russell, I. C. Geological History of Lake Lahontan. U. S. Geol. Survey, Monograph No. 11, 1885.
3. Jones, J. C. The Geologic History of Lake Lahontan. Carnegie Inst. Washington, Pub. No. 352, pp. 1–50, 1925.
4. Russell, I. C. Present and Extinct Lakes of Nevada. Nat. Geog. Soc. Monograph 1, No. 4, 101–136, 1895. Also in The Physiography of the United States (Nat. Geog. Soc.), 101–136, New York, American Book Co., 1896.
5. Louderback, G. D. Basin Range Structure in the Great Basin. Calif. Univ. Dept. Geol. Sci., Bull. Vol. 14, No. 10, pp. 329–376, 1923.
6. Jones, J. C. The Pleasant Valley, Nevada, Earthquake of October 2, 1915. Seism. Soc. Am., Bull. 5, 190–205, 1915.
Describes the earthquake rift visible from the airplane.
7. Chapman, R. H. The Deserts of Nevada and the Death Valley. Nat. Geog. Magazine, Vol. 17, 483–497, 1906.
8. Louderback, G. D. Basin Range Structure of the Humboldt Region. Geol. Soc. Am., Bull. 15, 289–346, 1904.
9. Louderback, G. D. General Geological Features of the Truckee Region East of the Sierra Nevada. Geol. Soc. Am., Bull. 18, 662–669, 1908.
10. Louderback, G. D. Lake Tahoe. Jour. Geog. 9, 277–279, 1911.

XII. **California.**

1. Lindgren, Waldemar. The Truckee Folio. U. S. Geol. Survey Folio No. 39, 1897.
2. Lindgren, Waldemar. The Colfax Folio. U. S. Geol. Survey Folio No. 66, 1900.
3. Lindgren, Waldemar. The Sacramento Folio. U. S. Geol. Survey Folio No. 5, 1894.
4. Lindgren, Waldemar. The Tertiary Gravels of the Sierra Nevada of California. U. S. Geol. Survey, Prof. Pap. 73, 1911.
5. Bryan, Kirk. Geology and Ground Water Resources of Sacramento Valley, California. U. S. Geol. Survey, Water Supply Paper No. 495, 1923.
6. Lawson, A. C., and others. The San Francisco Folio. U. S. Geol. Survey Folio 193, 1929.
7. Gilbert, G. K., and others. The San Francisco Earthquake and Fire of April 18, 1906. U. S. Geol. Survey, Bull. 324, 1907.
8. Willis, Bailey. Aerial Observation of Earthquake Rifts. Seism. Soc. Am., Bull., Vol. 11, No. 2, 136–139, 1921. Science, n.s., Vol. 54, 266–268, 1922.

INDEX

A star following the page number, thus 102*, indicates an illustration.

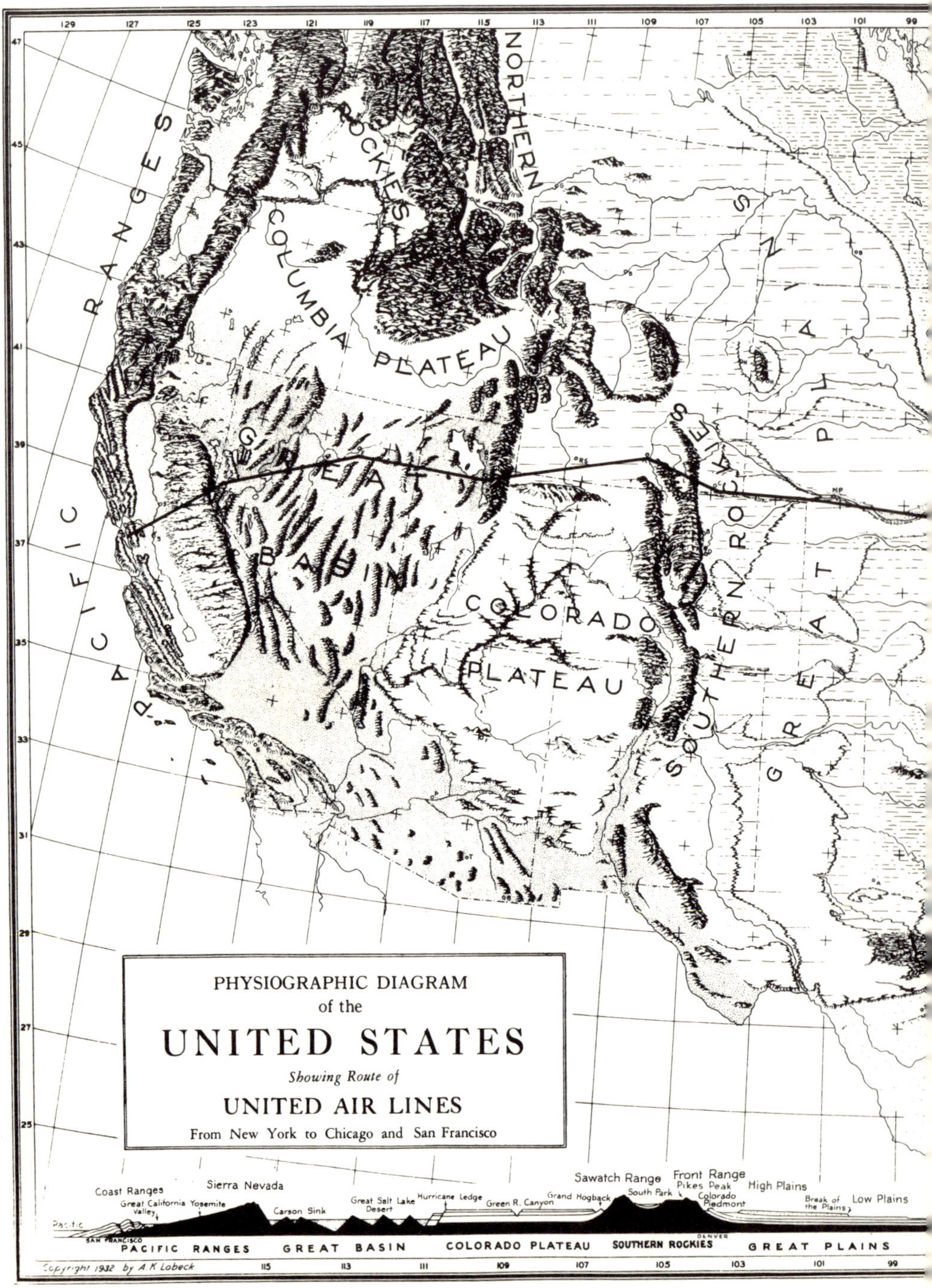
PACIFIC RANGES
NORTHERN ROCKIES
COLUMBIA PLATEAU
GREAT BASIN
COLORADO PLATEAU
SOUTHERN ROCKIES
GREAT PLAINS
PHYSIOGRAPHIC DIAGRAM
of the
UNITED STATES
Showing Route of
UNITED AIR LINES
From New York to Chicago and San Francisco
Coast Ranges
Great California Valley
Yosemite
Sierra Nevada
Carson Sink
Great Salt Lake Desert
Hurricane Ledge
Green R. Canyon
Grand Hogback
Sawatch Range
South Park
Front Range
Pikes Peak
Colorado Piedmont
High Plains
Break of the Plains
Low Plains
Pacific
SAN FRANCISCO
DENVER
PACIFIC RANGES
GREAT BASIN
COLORADO PLATEAU
SOUTHERN ROCKIES
GREAT PLAINS
Copyright 1932 by A.K Lobeck